FRESH WATER POLLUTION III: PROBLEMS & CONTROL

Papers by
John Cairns, Jr., A.W. Busch, E.L. Cronin, G.J. Kehrberger, K.D. Larson, Murray Stein, A.H. Cooke, J. David Eye, Kimrey D. Newlin, Douglas S. Diehl, James S. Mattson, Wilbur N. Torpey, C.R. Symons et al.

MSS Information Corporation
655 Madison Avenue, New York, N. Y. 10021

Library of Congress Cataloging in Publication Data
Main entry under title:

Fresh water pollution.

1. Water--Pollution. [DNLM: 1. Water pollution--Collected works. 2. Water pollution, Radioactive--Collected works. WA 689 F887]
TD420.F75 628.1'68'08s 72-13705

ISBN 0-8422-7103-1 (v. 3)

TABLE OF CONTENTS

CREDITS AND ACKNOWLEDGEMENTS

Busch, A.W., "Use and Abuse of Natural Water Systems," Reprinted with permission from *Journal of Water Pollution Control Federation,* 1971, 43:1480-1483.

Busch, A.W., "A Five-Minute Solution for Stream Assimilative Capacity," Reprinted with permission from *Journal of the Water Pollution Control Federation,* 1972, 44:1453-1456.

Cairns, John, Jr., "Thermal Pollution: A Cause for Concern," Reprinted with permission from *Journal of the Water Pollution Control Federation,* 1971, 43:55-66.

Cairns, John, Jr., "Ecological Management Problems Caused by Heated Waste Water Discharge into the Aquatic Environment," *Water Resources Bulletin,* 1970, 6:868-878.

Cairns, John, Jr., "New Concepts for Managing Aquatic Life Systems," Reprinted with permission from *Journal of the Water Pollution Control Federation,* 1970, 42:77-82.

Cairns, John, Jr.; Kenneth L. Dickson; and Albert Hendricks, "Pre- and Post-Construction Surveys," *Industrial Waste Engineering.*

Cooke, A.H., "Chlorination of Water as Carried Out by the Metropolitan Water Board," *Chemistry and Industry,* 1971, 6:164-169.

Coutant, C.C., "Thermal Pollution: Biological Effects," Reprinted with permission from *Journal of the Water Pollution Control Federation,* 1970, 42:1025-1027.

Cronin, E.L., "IV. Prevention and Monitoring: Pollution Prevention," *Proceedings of the Royal Society of London: Biology,* 1971, 177: 439-450.

Diehl, Douglas S.; Robert T. Denbo; Manmohan N; Bhatla; and William D. Sitman, "Effluent Quality Control at a Large Oil Refinery," Reprinted with permission from *Journal of the Water Pollution Control Federation,* 1971, 43:2254-2270.

Eye, J. David; and Lawrence Liu, "Treatment of Wastes from a Sole Leather Tannery," Reprinted with permission from *Journal of the Water Pollution Control Federation,* 1971, 43:2291-2303.

Kehrberger, G.J.; and A.W. Busch, "Mass-Transfer Effects in Maintaining Aerobic Conditions in Film-Flow Reactors," Reprinted with permission from *Journal of the Water Pollution Control Federation,* 1971, 43:1514-1527.

Larson, K.D.; R.E. Crowe; D.A. Maulwurf; and J.L. Witherow, "Use of Polyelectrolytes in Treatment of Combined Meat-Packing and Domestic Wastes," Reprinted with permission from *Journal of the Water Pollution Control Federation*, 1971, 43:2218-2228.

Mattson, James S.; and Frank W. Kennedy, "Evaluation Criteria for Granular Activated Carbons," Reprinted with permission from *Journal of the Water Pollution Control Federation*, 1971, 43:2210-2217.

Newlin, Kimrey D., "The Economic Feasibility of Treating Textile Wastes in Municipal Systems," Reprinted with permission from *Journal of the Water Pollution Control Federation*, 1971, 43:2195-2199.

Stein, Murray; Thomas C. McMahon; Allen S. Lavin; Frederick M. Feldman; Sanford R. Gail; and H. Edward Dunkelberger, Jr., "Enforcement in Water Pollution Control," Reprinted with permission from *Journal of the Water Pollution Control Federation*, 1971, 43: 179-199.

Symons, C.R., "Treatment of Cold-Mill Wastewater by Ultra-High-Rate Filtration," Reprinted with permission from *Journal of the Water Pollution Control Federation*, 1971, 43:2280-2286.

Torpey, Wilbur N.; H. Heukelekian; A. Joel Kaplovsky; and R. Epstein, "Rotating Disks with Biological Growths Prepare Wastewater for Disposal or Reuse," Reprinted with permission from *Journal of the Water Pollution Control Federation*, 1971, 43:2181-2188.

PREFACE

This three-volume collection on fresh water pollution is an addition to the MSS Topics in Ecology series.

Papers published from 1971-1972 are presented providing current information on new developments in major aspects of fresh water pollution including bacteriological, chemical, thermal and radioactive pollution. Application of the latest technical devices for the control of water pollution is also discussed.

Thermal Pollution

THERMAL POLLUTION–A CAUSE FOR CONCERN

John Cairns, Jr.

Society cannot continue to expand production of electric power and increase discharge of heated wastewater into the aquatic environment indefinitely without causing a major ecological crisis. Except in hydroelectric plants, the source of energy is locked in fossil fuel—coal, oil, or gas—or in nuclear fuel. This energy is transformed into heat by burning or by nuclear reactions. The heat then changes water into high pressure steam. As the steam turns the rotor in a turbine, the heat is transformed into mechanical energy. The turbine motor is connected with a generator where the mechanical energy is converted into electrical energy. But the amount of electrical energy that emerges from the plant is far less than the energy that went into it in the form of fuel. At present most of this heat is "thrown away," as with many other waste products, and it is capable of polluting the environment, just as surely as domestic wastewater, industrial waste, and agricultural waste. Some of the waste heat goes up the smokestack in hot gasses; some is disposed of in a nearby stream or lake where it may be a more serious problem. After the steam passes through the turbine, it goes into a condenser. There cooling water circulates through tubing and cools the steam until it condenses into water. It then can be returned to the boiler to begin the cycle all over again. Meanwhile, the cooling water is returned to the river, lake, or coastal waters of the sea from where it came, carrying the waste heat with it.

Like many other industrial wastes, this one was small at first (Figure 1). Early steam electric generating plants were much smaller than those of today. Less waste heat was discharged into waterways, and it could be dispersed quite rapidly. Its effects on the various forms of plant and animal life in the water were therefore possibly quite small. There were few studies made, so little is known about actual effects.

But today the rapid growth of population, the even more rapid growth in the demand for electric power, and the trend toward power grids rather than many small, self-contained electric systems are all leading in the direction of larger and larger generating plants. Fossil-fueled plants are now being built to produce 4 or 5 times the electricity of those built 20 yr ago. Further strengthening the trend to enormous generating capacity is the entrance of nuclear power into the field. Nuclear plants are not yet economically feasible in small sizes. Nuclear generating plants are therefore built to produce even more electricity than the new conventional fossil fuel plants. They also produce electricity less efficiently. That is, less of the energy is transformed into electricity; more of it becomes waste heat. Nuclear energy power generation requires about 50 percent more cooling water than com-

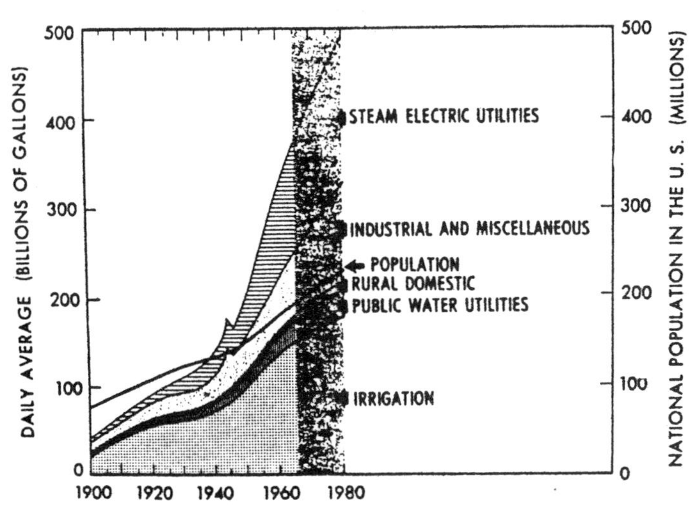

FIGURE 1.—Water use in the United States. [Based on Picton, W. L., "Water Use in the United States." (1960).]

parable power output by conventional methods (1).

Water Balance

The total average rainfall in the U. S. is 30 in./yr (76 cm/yr) of which about 70 percent is returned to the atmosphere through evaporation and from trees and other plants that lose water to the atmosphere by a process called transpiration. This leaves about 1,200 bil gpd (4,540 mil cu m/day) for surface and subsurface water supplies. Singer estimates that steam-electric power generators will pass 25 percent of this through their cooling systems by 1985 (2). A single power plant now in existence requires up to 0.5 mil gpm (1,892 cu m/min) for cooling purposes; running at full capacity around the clock, this would mean 720 mgd (2,725,000 cu m/day).

Some of this is "consumed." That is, it evaporates and is lost, at least temporarily, to other uses and users. The amount "consumed" has been estimated as averaging 20 acre-ft (24,700 cu m) for each megawatt of electricity generated (3). A large (1,000-megawatt) plant would consume 20,000 acre/ft/yr (24.7 mil cu m/yr) of water. Most of the cooling water, however, is not consumed, but is returned—at a higher temperature—to the waterway from which it was drawn.

Unfortunately, the distribution of water is geographically uneven and it also varies seasonally. In certain watersheds, a major portion of the water available may be required by the local power plant during low flow periods. To make matters worse, this period of maximum use may coincide with the warmest period of the year when surface waters are already at or near maximum temperatures. For example, the Monticello nuclear generating plant (without cooling towers) would use about 65 percent of the entire flow of the upper Mississippi River at times and could raise the river's temperature in that region as much as 16°F (≅9°C). How far the water would flow before returning to its normal temperature is not known (4).

As the cooling water returns to the stream or other source, it may carry with it very small amounts of heavy metals from corrosion of the tubes in the condenser or chlorine which is sometimes used to clean those tubes. Minute aquatic organisms, drawn through the system along with the water, have been subjected to increased temperatures and may be affected. Water drawn from an estuary, for example, may include fish eggs and larvae, and the microscopic plants (phytoplankton) and animals (zooplankton) that abound in these waters. Some investigators feel that few of these will survive the shock of passing through the condenser (5). Unfortunately, little research has been carried out on organisms drawn through cooling systems, and results of the research that has been done are not in agreement. Even if all the organisms that pass through the condenser are killed, the effect on the ecosystem is difficult to predict. It would probably vary from one area to another. Preliminary results from research in the author's laboratory indicate that these effects are not particularly severe.

The effect of the heat is not limited to the water actually drawn through the generating plant and the forms of life contained in that water. When the heated water returns after it has flowed through the power plant, the entire aquatic ecosystem—the interrelated system of living things in the larger body of water—may be affected even though only a portion of the water was used. This added heat may then change the character of both the system of living things and the receiving water, which together make up the aquatic ecosystem.

Man is a simplifier of complex ecosystems and a creator of simple ecosystems. What makes a natural ecosystem is the multitude of interactions such as predation, parasitism, and competition, which occur between the many species of organisms, both plant and animal, that make up the biological part of an ecosystem. Seasonal changes occur, and species have good periods and bad periods, but over the long run the important thing is the relative constancy in number of species present and in their relationships. This functional stability is the mark of a healthy natural ecosystem.*

Corn fields and fermentation vats are good examples of simplified ecosystems. These are notoriously unstable and require constant care and maintenance. Large amounts of pesticides are used each year to keep insect populations at low levels in the cornfield, and brewers must carefully measure their ingredients and control the environmental conditions so that the desired end-product is obtained. Simple ecosystems are not necessarily "bad" nor are complex systems necessarily "good," but they do have different characteristics.

Environmental Stresses

Each technological advance has produced increased stress on the natural environment. As stress increases and more and more species are eliminated, the many biological interactions that maintain ecosystem stability are short-circuited. "Weed" species (those able to tolerate the stress) flourish as their more sensitive competitors and predators are removed. With the sudden application of a stress factor to an ecosystem, only those species with a preexisting ability to cope with the new conditions will survive. Populations can accommodate to gradual change (i.e., spread out over many generations) by the process of natural selection. In fact, that is how species and ecosystems evolve.

Even with a relatively rapid change in conditions, species are rarely eliminated *en mass*. As stress is applied to an ecosystem, the first response is a reduction in the number of individuals of the more sensitive species. At the

* This explanation is necessarily oversimplified. For readers who would like to learn more about ecosystems, "Ecology" by Eugene F. Odum, published by Holt, Rinehart and Winston, New York, N. Y. (1963), is recommended.

same time, numbers of individuals of the more tolerant species may increase so that the total number of individuals has hardly changed. If the stress increases, species begin to disappear as conditions exceed their tolerance levels. For example, in a small watershed in eastern Pennsylvania, 23 species of fish were found in the main stream. In a polluted tributary three pollution-resistant species, goldfish, creek shiners, and killifish, were the only species found. The total pounds per acre of fish in the clean water stream and in the polluted tributary were quite comparable. The three pollution-tolerant species can survive and reproduce in the stressed environment free from competition with or predation from the other 20 species (6). This is typical of polluted waters, whether the pollution is from heat or from other forms of waste.

Although in terms of geologic time, man's modification of the environment must be considered sudden, to an observer trying to document the changes as they occur, it is a very gradual process. The replacement of individuals of sensitive species with those of more tolerant species is also a biological accommodation that masks change. Such a change is often not obvious even to a trained observer unless he has studied the area sufficiently well over a period of time to have statistically reliable data. Eventually, if an increase in stress continues, a point will be reached where the ecosystem collapses. The fish are gone, the water smells, industrial and municipal wastes are not degraded as well as they formerly were, and one does not need to be an ecologist to know that things have changed.

Discharge of heated wastewater seems to have the same effects as other types of stress in that there are three levels of biological response:

1. A range within which no abnormal response to change is noted;

2. A zone of graded response—increased stress produces increased response (more dead fish and fewer species);

3. A threshold beyond which further increases elicit no further response (because the system is incapable of further response—not because further increase is suddenly harmless).

Most aquatic organisms can withstand the effects of heat only within very narrow limits. They are further handicapped by not having regulatory mechanisms (such as sweating) which buffer effects of increased heat in the environment. It is true that some can move to cooler areas, but this is avoidance rather than regulation and is possible only when a cooler area exists. Moreover, the discharge of heated water into a stream can create a hot water barrier which effectively blocks the spawning migrations of many species of fish, as well as other, less spectacular movements. The temperature an aquatic animal can tolerate depends greatly on the conditions in the normal environment—tropical fishes are adapted to warm waters that would kill a brook trout. Sometimes in areas with very stable temperatures the margin of safety is very slight. The polyps that build coral reefs are killed by temperatures only 2° or 3°F (2.2° to 6.4°C) above those at which they carry out a normal existence.

Though abrupt application of heat can kill bacteria and other microorganisms (pasteurization is an example) the probable result of gradual changes in a natural environment is exclusion of nontolerant species by resistant species. That is, those species more tolerant of the new conditions outperform the others and reproduce more rapidly while the less tolerant reproduce slowly or not at all. The result is a qualitative shift in kinds of species present which may or may not be accompanied by a change in total numbers of individuals. Diatoms, green algae, and blue-green algae, each group including a number of species, may all be found together in an aquatic ecosystem.

However, Figure 2 shows how the species relationships, in one experiment, shifted as the temperature rose from 68°F (20°C) to 104°F (40°C). The number of species of diatoms dropped sharply; the trend among blue-greens was the opposite—there were more species present in the warmer water. Species of greens increased with rising temperature until it passed 86°F (30°C) and then dropped off again. Present evidence suggests that blue-greens are not as suitable as diatoms to the organisms dependent on algae as a food supply. Many blue-greens produce unpleasant odors, and some are toxic to shellfish and other organisms.

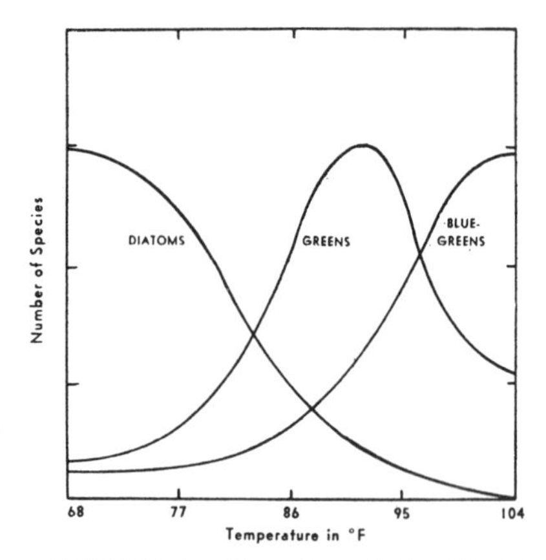

FIGURE 2.—How the relationship between three groups of algae shifts as the water temperature rises. [(°F − 32) 0.555 = °C.]

The major effects of heat pollution on higher aquatic organisms are:

1. Death through direct effects of heat;
2. Internal functional aberrations (changes in respiration, growth);
3. Death through indirect effects of heat (reduced oxygen, disruption of food supply, decreased resistance to toxic substances);
4. Interference with spawning or other critical activities in the life cycle; and
5. Competitive replacement by more tolerant species as a result of the above physiological effects.

Experiments suggest that some aquatic organisms can become acclimated to higher temperatures than those to

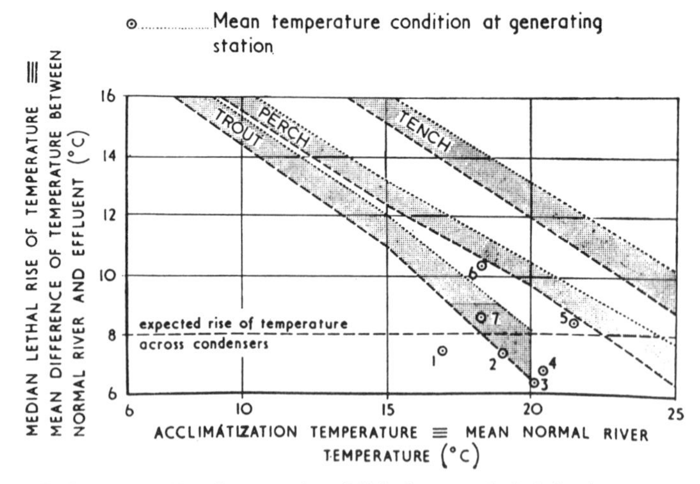

FIGURE 3.—For three species of fish the mean lethal rise in temperature decreases as acclimatization temperature increases. [(°F − 32) 0.555 = °C.]

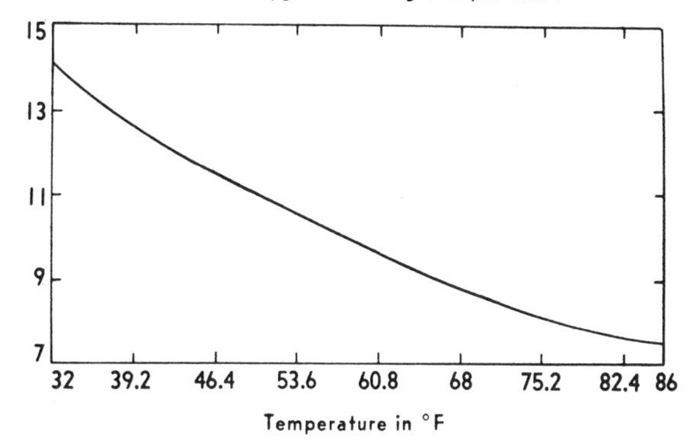

FIGURE 4.—As water is heated, its ability to hold dissolved oxygen (DO) decreases, making less oxygen available for respiration by aquatic organisms. [°C = 0.555 (°F − 32).]

which they are normally accustomed. But even in the laboratory, gradual acclimation to much higher temperatures has limitations. The experiments of Alabaster (7) have shown that for three species of fish, the mean lethal rise in temperature decreases as acclimatization temperature increases (Figure 3). Of course, one should be quite cautious about using results even from carefully controlled laboratory experiments to develop standards for use in natural environments. Some organisms survive higher temperatures in natural situations than in laboratory tests. The reverse may also be true. In the laboratory, an organism may have near optimal conditions with the exception of the changing temperature being studied. Many organisms in the natural environment already live under less than optimal conditions and are often subject to other kinds of stress at the same time the temperature is changing. Another form of stress is a secondary effect of heated water. Heating water decreases the solubility of oxygen (Figure 4) and, within certain limits, increases bacterial activity. These events lower the amount of oxygen available to the higher aquatic organisms.

The problems caused by heated wastewater discharge into lakes probably are as complex as those in rivers, though data on lakes is quite scarce. Shallow lakes are mixed by the wind and may therefore have little temperature variation throughout, but deep lakes are quite different. The most dramatic ecological events for deep lakes in the temperate zone are the seasonal turnovers. In the spring, surface waters warm to 39°F (4°C) (the temperature at which water has its greatest density) and sink to the bottom, bringing the nutrient-rich bottom water to the surface. In the fall, the surface waters cool, and again a turnover occurs. The algal blooms which often follow these turnovers are well known. The ecology of temperate zone lakes is largely determined by turnovers and by the stratification of the water into layers with varying temperatures during the intervening periods. The chemical, physical, and biological structure of lake ecosystems is keyed to these events. A power plant that places a layer of warm water on a lake surface may disrupt the circulation pattern and prevent turnovers from occurring or change the season at which they take place. The effects of this on lake ecology might be drastic and possibly disastrous.

When the New York State Electric and Gas Corporation proposed construction of a nuclear generating plant on Lake Cayuga, near Ithaca, N. Y., a group of Cornell scientists pointed out that during the summer, the plant would withdraw from the lower, cooler layer of the lake roughly 10 percent of its average volume. Seven hundred seventy mgd (2.9 mil cu m/day) of water heated to 65° to 70°F (18° to 21°C) would be added to the top, warmer layer. The scientists predicted that this would substantially delay the usual cooling and mixing of the lake in the fall. The addition of heated water in the winter might cause the summer stratification to begin earlier. The lake's growing season would therefore be extended at both ends, with a higher rate of biological production (8). In other words, discharge of heated wastewater might increase the growth of algae and speed the eutrophication or aging process in the lake. It should be noted that these comments apply primarily to smaller lakes. The Great Lakes pose a unique situation that will require particular attention and probably a set of standards that are different from those of ordinary lakes.

One might think that the vast oceans would be immune to effects of heat pollution and that therefore electric generating plants could best be placed along the coasts, but since marine life is frequently concentrated in coastal areas, this may not be the easy solution it seems to be on first inspection. For example, it has recently been reported that a combination of heat and salt pollution caused by waste discharge from the Point Loma saline water conversion plant near San Diego, Calif., has affected intertidal marine life (9). Temperatures taken at a power station in South Wales ranged from about 73°F (22°C) in winter to about 99°F (37°C) in summer in an estuary where mean sea temperatures range from about 45° to 63°F (7° to 17°C) (10).

Historically, it has been general practice to assess water pollution in purely chemical terms and to assume that if one stays within certain arbitrary limits, no harmful biological effects will occur. Frequently, little or no effort is spent directly assessing the biological effects. Historic reasons for this seem to be that chemical data were comparatively easily acquired and could be gathered more rapidly. Acquiring biological data takes far less time today because many improved biological assessment methods are available. However, the most compelling reason for using biological assessment methods is that knowing the chemical or physical characteristics of a waste or receiving water makes possible only an estimate of the effects on living organisms. These effects may be accurately determined only by a direct biological assessment. Biological study seeks to understand environmental interactions, but it does not indicate the exact substances or conditions involved. Rather than rely on any single approach, biological, chemical, and physical assessments should all be used in studying pollution.

Initial attempts to avoid biological problems resulting from the discharge of heated wastewater were primarily directed toward avoiding lethal temperatures. However, because temperature changes often act to time the occurrence of biological events, it is quite possible to disrupt biological systems fundamentally while staying within nonlethal temperatures by changing the pattern of temperature change. Gardeners know that certain bulbs and seeds require cooling in order to produce plants. The temperature pattern during the cooler months also affects the life cycles of many aquatic organisms so that where heated water is discharged, the receiving water may not get cold enough for many species. The emergence pattern and reproductive cycle of some aquatic insects depends on seasonal temperature cycles. A population of insects emerging in

March when it would normally appear in May would surely meet with disaster. Further, its absence in May would be likely to affect other animals that depend on it for food. A whole chain of events that would seriously disrupt normal biological interactions could be set into motion by the mistiming of seasonal temperature changes*.

The related and complementary problems of increased energy production and population growth are forcing man to make a choice between a complex environment with considerable functional stability and a simplified environment with increased management costs. A good example of a stable environment is a complex forest consisting of a great variety of plants and animals that will persist year in and year out with no interference from man. This ecosystem is a complex mixture of biological, chemical, and physical interactions, many of which cancel each other out. For example, if there are several predators regulating a population of rodents and one disappears, the effect often is reduced by a population expansion of other predators also feeding on the rodents. Therefore, the system is one of dynamic equilibrium, with the system itself being stable but with many of its components undergoing change. A simple system such as a corn field produces a quantity of material immediately useful to man but is notoriously unstable. Without constant care and attention it would cease to be useful and would disappear. So the history of civilization has been one of widespread simplification of the environment with consequent increased management requirements.

* Documenting the effects of temperature in the aquatic environment in detail is clearly outside the scope of this article. Kennedy and Mihursky (11) recently prepared a superb bibliography of 1,220 key references on this subject. The increase in literature on temperature effects in recent years is quite dramatic. More papers on this subject were produced in a single year in the sixties than in all the years from 1900 to 1920. Raney and Menzel (12) have an extremely useful 34-page bibliography primarily devoted to the effects of heated discharges on fish. For those interested in the physiology of temperature adaptation, "Molecular Mechanisms of Temperature Adaptation" by Prosser (13) will provide an excellent starting point. A general discussion of all phases together with an impressive selected bibliography are included in "Temperature and Aquatic Life" by Holdaway *et al.* (14). The excellent pair of books edited by Parker and Krenkel (15) furnish much valuable information on heated wastewater problems. Anyone interested in further information may consult these source materials.

The Effects of Technology

When man simplifies an ecosystem, he creates numerous ecological problems. When a complex natural area is cleared and planted with corn, a single type of food is concentrated in a limited area and it requires protection against insects and other pests. To reduce the number of these pests, pesticides are applied and the diversity of life in the soil and in the field is further reduced. Because certain of the pests may become resistant to the insecticides after repeated application, there is a gradual escalation of concentrations of these which may further simplify the ecosystem. At the same time, other organisms including man are beginning to have substantial amounts of these pesticides incorporated into their own tissues. For example, concentrations of DDT in the fat deposits of human tissues in the U. S. average 11 ppm and Israelis have been found to have as much as 19.2 ppm (16). So the overall question is whether not only thermal pollution but also all the other problems producing the environmental crisis can be controlled. Gershinowitz (17) summed up the problem rather succinctly, "As soon as one becomes involved in any one of the specific problems of the pollution of the environment it becomes apparent that no one problem can be treated in isolation. Methods for waste disposal, whether they are concerned with gases, liquids or solids, interact with each other; incineration of solids can cause contamination of the air, sanitary fills can cause contamination of water supplies, substitutes for incineration, such as maceration, can increase the load on water purification systems." To create a workable method for a single form of stress without considering the others will only cause a slight delay in the inevitable catastrophe. Only with total environmental planning, including population control, will a meaningful program to insure a harmonious relationship with the environment be possible.

The carrying capacity of the earth usually has been estimated by calculating the number of people and the food resources to feed them at certain nutritional levels. In short, food supply was thought to be the ultimate limitation on the size of the human population. However, it may well be that the ultimate limitation on the number of people the earth can support will be the capacity of the environment to transform wastes into acceptable or useful materials. Somehow the cyclic nature of useful things has been forgotten, possibly because man talks of a consumer society. Actually, it is a user society which temporarily uses consumer goods and then discards them in forms such as domestic wastewater, carbon dioxide, and tin cans into the environment. Because man depends on the environment as a life support system, an attempt should be made to understand its waste transforming capacities which are as integral a part of the life-support system as is the food supply. Man can no longer afford merely to insure the survival of the ecosystems which provide oxygen and transform wastes. He must also determine the conditions required for optimal function, because, with the predicted increase in rate of loading with wastes, nothing less than optimal function will prevent disaster.

Priorities

There are four basic alternatives that are open with regard to the heated wastewater problem, which may be chosen singly or in various combinations:

1. Placing all heated wastewater in streams, lakes, and oceans without regard to the effect and considering the environmental damage as a necessary consequence of the increased power demand;
2. Using, but not abusing, present ecosystems. This means regulating the heated wastewater discharge to fit the receiving capacity of the ecosystem;
3. Finding alternative ways to dissipate or beneficially use waste heat; and
4. Modifying ecosystems to fit the new temperature conditions.

Using, but not abusing, present ecosystems would entail determining the receiving capacity of each ecosystem for all the wastes being discharged into it. This would include insecticide runoff, domestic wastewater, silt from road building, and heavy metals, as well as heated wastewater. Without a doubt, the receiving capacity of an ecosystem would vary seasonally. Therefore, adjusting waste discharge to fit the needs of the system would require either (*a*) limitation of the population at something near its present size and improving waste treatment facilities in areas where serious environmental degradation has occurred, or (*b*) if the population is to increase, toleration of a lower quality environment or drastically changed waste disposal practices. Both of these are going to be quite difficult to achieve and may be virtually impossible under present circumstances. There are two positions taken by a few power company spokesmen that are outdated in today's complex society in which single purpose use of a major natural resource is virtually impossible:

1. The company has a legal obligation to supply safe and adequate elec-

tric service and any attempt to regulate wastes from this process is contrary to public interest;

2. Criteria for controlling heated wastewater discharge are needlessly restrictive and not based on scientific fact.

There are major weaknesses in these positions that fortunately are realized by many electric power company spokesmen. The first weakness is that the human race depends not only on electricity but on the total ecosystem including the water, the soil, and the air. The second point that the criteria are not based on sufficient scientific fact probably means that biologists cannot tell the precise death point for each and every species in the environment. This is true and will probably remain true for generations to come; however, it is known that substantial changes in temperature will have major effects on the functioning of an ecosystem including its ability to transform and degrade wastes. One should remember to distinguish between standards and criteria. Standards can be changed without changing criteria.

On the other hand, if a power company has demonstrated that the ecological quality of an area will not be degraded by the proposed heated wastewater discharge and is prepared to provide an ecological monitoring program to show that environmental quality is being maintained, requiring further treatment for its own sake seems irrational. In essence, this latter requirement would mean applying economic resources where no significant results will be produced, and very likely diverting resources from other areas where they are badly needed. The basic question that society faces is what are the most beneficial uses of a nonexpandable resource, the environment, which will permit it to survive as a vigorously functioning, dependable system available for future beneficial uses. Neither conservationists who refuse to recognize the need for rational industrial use of natural resources nor industrial representatives who feel that industrial use of the environment should never be questioned are contributing to the solution of a common pressing problem. Both positions are untenable in the present circumstances. Unrestricted industrial use would so damage the ecosystems that the life support capacity would be either destroyed or seriously degraded. The extreme conservationists' attitude would not permit the type of industrial society to which the U. S. is accustomed. If the kind of harmonious working relationship that will permit full beneficial use of the environment is to be achieved, less extreme positions will have to be taken by spokesmen on both sides.

The third alternative—that of finding other ways of dissipating heat or using it for other beneficial purposes, such as heating greenhouses in the north—deserves more attention than it has received in the past. One alternative to the open-cycle system (in which the water is passed through the condensing units and returned to the stream or lake from which it was taken at a considerably higher temperature) is the use of cooling towers. Wet-type hyperbolic cooling towers have been used in Europe for the past 50 yr and have been used widely enough in the U. S. so that their operational characteristics are fairly well understood. Dry-type cooling towers are a relatively recent development and are less well understood. The first dry-type tower was erected at the Rugeley Station in Great Britain serving a unit rated at 120,000 kw and went into service in December 1961. There seems to be no question even from the relatively small amount of information available on dry-type cooling towers that the cost of operating these will be considerably greater than the wet towers; estimates run as high as four times as much. On the other hand, the wet-type cooling towers do cause a considerable loss of water through evaporation, and makeup water is required periodically as well as the discharge of the nonevaporative portion of the water used in cooling which would contain a rather large dissolved solids load. An aesthetic objection might also be raised to the use of all cooling towers because for some typical power generating units now being constructed, the requirement would be five wet-type hyperbolic towers, each tower being several hundred feet high and several hundred feet in diameter. Obviously in most areas these would be difficult to hide. It is quite evident that some means of making social decisions, which include the aesthetic as well as the other factors, must be developed. One of the most common objections to the installation of cooling towers is that they are quite expensive and that the public is unwilling to pay the additional cost for electric power. However, the results of a recent Gallup poll (18) indicate that this may be a fallacious generalization. About half (51 percent) of all persons interviewed said they are "deeply concerned" about the effect of air and water pollution, soil erosion, and destruction of wildlife. About one-third (35 percent) said they are "somewhat concerned." Only 12 percent said they are "not very concerned." Even more important, about three out of every four people interviewed backed up their concern by stating that they would be willing to pay additional taxes to improve natural surroundings. Because people are never eager to pay additional taxes, these figures indicate a strong desire on the part of the American people to protect the environment and a willingness to pay directly for the cost of this protection. Senator Henry M. Jackson of Washington, Chairman of the Senate Interior Committee, summed up this viewpoint as follows: "A new attitude of concern for values which cannot be translated into the language of the market place or computed in cost-benefit ratios is being felt and seen in citizen efforts to save parks, open spaces, and natural beauty from freeways, reservoirs, and industry." Senator Jackson further stated, "People are no longer complacent about the quality of their surroundings, about the use of the environment, and the way public resources are being administered." It seems quite clear that there is both public interest in the development of alternative uses of heated discharge waters (such as heating greenhouses) and a willingness to pay for treatment that would not introduce dangerous quantities of heated wastewater into the natural environment. Man depends on a life support system that is partly industrial and partly ecological. Unfortunately, he has reached a stage of development where the nonexpandable portion of the life support system, the ecological part of the environment, is endangered by the expanding industrial portion of the environment. Developing a balanced life support system that serves the greatest variety of beneficial uses will not be possible unless narrow discipline-oriented, single-purpose views toward environmental management and use are abandoned. Optimal use and management of such a complex system will only be possible with the full cooperation of engineers, ecologists, congressmen, economists, urban and regional planners, regulatory agencies, businessmen, geologists, to mention just a few of the many sources of critical points of view. The decisions are too important to be left to a single group—too complex to be solved by a single discipline.

Finally, waste discharge regulations that cause needless expense without protecting the receiving ecosystem must be revised. Failure to do so will undoubtedly alienate industrial representatives and divert energy and money from areas where improved waste control practices will have marked benefits.

Acknowledgments

The author is grateful to *Scientists & Citizen* (now *Environment*) for giving permission to quote parts of the author's article "We're in Hot Water" which appeared in Vol. 10, No. 8, pp. 187–198, 1968.

References

1. Kolflat, T., Testimony in "Thermal Pollution—1968." Hearings before the Subcommittee on Air and Water Pollution of the Committee on Public Works, U. S. Senate, 90th Congress, 2nd Session, U. S. Govt. Printing Office, Washington, D. C., 63 (Feb. 1968).
2. Singer, S. F., "Waste Heat Management." *Science,* **159**, 3820, 1184 (1968).
3. Bennett, N. B., Jr., Address to Western Water and Power Symposium, Los Angeles, Calif., April 9, 1968.
4. Abrahamson, D. E., and Pogue, R. E., "Some Concerns about the Environmental Impact of a Growing Nuclear Power Industry: I. The Discharge of Radioactive and Thermal Wastes." *Jour. Minn. Acad. Sci.* (In press.)
5. Mihursky, J., Testimony in "Thermal Pollution—1968." Hearings before the Subcommittee on Air and Water Pollution of the Committee on Public Works, U. S. Senate, 90th Congress, 2nd Session, U. S. Govt. Printing Office, 102 (Feb. 1968).
6. Trembley, F. J., Testimony in "Thermal Pollution—1968." Hearings before the Subcommittee on Air and Water Pollution of the Committee on Public Works, U. S. Senate, 90th Congress, 2nd Session, U. S. Govt. Printing Office, 94 (Feb. 1968).
7. Alabaster, J. S., "Effects of Heated Effluents on Fish." *Air & Water Poll.,* **7**, 541 (1963).
8. Arnold, D. E., *et al.,* "Thermal Pollution of Cayuga Lake by a Proposed Power Plant." Citizens Committee to Save Cayuga Lake, Ithaca, N. Y. (1968).
9. Leighton, D., *et al.,* "Effects of Waste Discharge from Point Loma Saline Water Conversion Plant on Inter-Tidal Marine Life." *Jour. Water Poll. Control Fed.,* **29**, 1190 (1967).
10. Naylor, E., "Biological Effect of a Heated Effluent in Docks at Swansea, South Wales." *Proc. Zoological Society of London,* **114**, 253 (1965).
11. Kennedy, V. S., and Mihursky, J. A., "Bibliography on the Effects of Temperature in the Aquatic Environment." Natural Resources Inst., Univ. Maryland, College Park, mimeo, (1967). Reprinted in "Thermal Pollution—1968." Hearings before the Subcommittee on Air and Water Pollution of the Committee on Public Works, U. S. Senate, 90th Congress, 2nd Session, U. S. Govt. Printing Office (1968).
12. Raney, E. C., and Menzel, B. W., "A Bibliography: Heated Effluents and Effects on Aquatic Life with Emphasis on Fishes." Fernow Hall, Cornell Univ., Ithaca, N. Y. (1967).
13. Prosser, C. L., "Molecular Mechanisms of Temperature Adaptation." AAAS, Pub. No. 84, 390 pp. (1967).
14. Holdaway, J. L., *et al.,* "Temperature and Aquatic Life." Tech. Advisory and Investigations Branch, Tech. Serv. Progr., FWPCA, U. S. Dept of Int., Lab. Investigations Series, No. 6, Washington, D. C. (1967).
15. Krenkel, P. A., and Parker, F. L., "Biological Aspects of Thermal Pollution." Vanderbilt Univ. Press, Nashville, Tenn. (1969) and Parker, F. L., and Krenkel, P. A., "Engineering Aspects of Thermal Pollution." Vanderbilt Univ. Press, Nashville, Tenn. (1969).
16. Ehrlich, P. B., "The Population Bomb." Ballantine Books, New York, N. Y. (1968).
17. Gershinowitz, H., "The Environmental Studies Board." Presented at the Annual Meeting of the National Research Council in Washington, D. C. (March 1969).
18. Cahn, R., *Christian Science Monitor,* 5 (March 11, 1969).

Additional Reference

Raney, E. C., and Menzel, B. W., "Heated Effluents and Effects on Aquatic Life with Emphasis on Fishes." Cornell Univ., Water Resources and Marine Sciences Center and Ichtyological Associates, Ithaca, N. Y., Bull. No. 2, 470 pp. (1969).

Thermal Pollution—Biological Effects

Reviewed by C. C. Coutant

Biological effects of thermal discharges to aquatic ecosystems are many and diverse. They may range from direct lethal effects of high temperatures to subtle changes in behavior, metabolism, performance, community structure, food-chain relationships, and genetic selection. These effects may be studied both in the waste discharge area of a power station and in the physiology laboratory, with each approach providing valuable insights into this increasingly important aspect of environmental attention.

Reviews

A number of reviews published in 1969 dealt with temperature effects in aquatic ecosystems. Some were directed toward practical problems associated with industrial thermal discharges, while others dealt with basic questions of subtle responses of organisms to temperature changes.

Clark (40) reviewed the general problems of thermal pollution and aquatic life from the viewpoint of a fisheries scientist. Information was given on temperature measurement by infrared imagery, thermal patterns at power-plant discharges, thermal tolerances and preferences of common fish and food-chain organisms, and meth ods of reducing heat loading of aquatic environments.

The U. S. Federal Power Commission (189) prepared a staff report that discussed environmental problems in the disposal of waste heat from steam-electric plants. Topics included were effects of heat on aquatic life, potential beneficial uses of waste heat, state and federal water quality standards, some current research programs, and apparent research needs, including studies on site selection.

Conditions for coexistence of aquatic communities with the expanding nuclear power industry were discussed by Davis (47), who considered water quality controls, thermal effects on aquatic communities, industrial heat discharges, and diversion of cooling water from the aquatic environment, as well as radioactive and chemical emissions.

Two papers by engineering specialists stressed the need for advance planning in selecting power reactor sites in order to provide adequate consideration of ecological factors, particularly those factors involved in disposal of waste heat. Jaske (82) stressed the desirability of determining the cumulative cooling capacities of regional systems, e.g. upper Mississippi River, based on system models of thermal additions. Morgan and Bremer (126) approached the problem at specific locations and stressed the need for ecological surveys prior to site selection and also before and during plant operation.

The 1968 literature pertinent to assessing the effects of thermal discharges into aquatic environments was reviewed by Coutant (43). A total of 111 articles were abstracted to indicate the nature of their scope rather than to discuss their individual importance.

Raney and Menzel (151) published a bibliography containing 1,870 references on heated effluents and their effects on aquatic life, particularly fish. All titles were permuted and indexed alphabetically on each significant or key word. The bibliography also included the standard author and title index.

The European Inland Fisheries Advisory Commission (57) reviewed literature on thermal effects on fish, principally from eastern Europe, and presented recommendations for permissible maximum temperatures and seasonal increments of temperature. The review made available Slavonic literature virtually unknown to the rest of the world.

Banks (13) reviewed literature on the upstream migration of adult salmonids and included a section on temperature effects. He concluded that evidence of temperature influence on upstream migration was both conflicting and inconclusive. Although adult salmonids have preferred temperatures, these may be unrelated to initiation or blockage of migration, depending on the locale studied.

Temperature adaptations of amphibian embryos were reviewed by Bachmann (10) with generalizations that may apply to other aquatic vertebrates. A simple equation was presented to describe the development rate of embryos between any two stages at various temperatures. The equation yielded values for two constants describing the adaptive temperature and the relative time for development at that temperature for every species.

The heat and cold resistance of free-living and endoparasitic species of protozoa, and the effects of ecological factors (i.e., salinity and food) on temperature adaptations, were discussed in a Russian text by Sukhanova (177). Information was given on the mechanisms of thermal effects in cells, life-cycle differences in thermal adaptations, and the importance of temperature adaptation to protozoan evolution.

The proceedings of a National Symposium on Thermal Pollution were published in two volumes, one principally on biological aspects (97) and the other principally on physical and engineering aspects (138). Each volume included several reports presented by people with administrative or research responsibilities bordering on their respective topics. Biological reports from these proceedings are included in this review.

Patrick (140) reviewed thermal effects on freshwater algae. There are thermophilic, kryophilic, and mesophilic species of algae, and a single species may have strains with very much different thermal optima. Many laboratory experiments have been conducted to determine the effects of increased temperature on cell growth, photosynthesis, and metabolic functions. Additional studies have been conducted in artificial or natural stream environments. There are comparatively few detailed evaluations of the effects of thermal discharges on algae and these studies were presented. Round (156) briefly considered the interactions of light and temperature on various aspects of the ecology of algae.

Hedgepeth and Gonor (74) reviewed the effects of temperature changes on marine benthos, especially inshore organisms. Their principal concerns regarding the prediction of temperature-induced changes in the fauna were (*a*) the inadequacy of most available temperature measurements of organisms in their normal habitat (e.g., they found that temperatures of mussels exposed at low tide normally exceeded 25°C, while the adjacent water was 10° to 14°C); (*b*) a disproportionate emphasis on potential lethal effects to adults while ignoring subtle, long-term temperature changes important to population dynamics; and (*c*) the lack of recognition that temperature fluctuations are not just tolerated by organisms but that they are often necessary, especially for reproductive cycles. Some examples of site studies and of marine aquiculture in heated waters were given critical appraisal. In formal discussion of this paper, North (135) stressed the urgency of establishing general ecological guidelines as aids in designing and locating thermal discharge systems. North presented, in outline, a guide to thermal discharge siting. In these considerations, direct physiological damages from heat may be minor, but alterations in ecological relationships within communities may be great.

deSylva (178) gave a comprehensive review of the effects of heated effluents on marine fish that emphasized the complex and interrelated events resulting from even a slight temperature increase. deSylva concluded that these complex effects were not understood and could not be predicted with confidence. The need was stressed for careful surveys and reasonably detailed knowledge of the environment prior to selection of sites where thermal pollution might occur.

Thermal variability in natural environments and the normal adaptations of aquatic organisms in temperature zones to this variability were stressed by Wurtz (199), who also discussed heat and freshwater benthos. Some adaptations of benthic organisms were believed by some individuals to occur through physiological mechanisms still imperfectly known. The community provided further adaptation by changing species composition to provide active biota in even the most altered aquatic environment. Cairns (34), who discussed this paper, stressed the opposite opinion that environmental alterations such as thermal discharges should not disrupt the stable ecosystem of the receiving water although some biota, with a degree of community complexity, might arise under the new conditions. Preliminary ecological surveys followed by conscientious plant location and operation were observed to have reached this goal in many instances.

Ecological changes of applied significance induced by the discharge of heated waters were discussed by Hawkes (72). He reviewed relevant ecological principles, temperature as a natural ecological factor, and field and laboratory studies of ecological and chemical changes caused by artificial rises in water temperatures. In formal discussion of this paper, Welch (195) described some observed and potential biological effects of heated discharges in the Tennessee Valley. Problems were emphasized in the interpretation of data in terms of detrimental or beneficial effects.

Research needs for thermal pollution control were described by Allen (5).

She believed that field surveys should be adapted to the unique conditions of a given site. Information on the effects of temperature, reproduction, and growth of desired species was especially lacking, followed by similar data on food-chain organisms in all their life stages. Synergistic and long-term effects are poorly known, especially as interacting parameters such as salinity are altered by thermal discharges. Relationships of thermal effects to beneficial and nonbeneficial water uses should be stressed in pollution control and attendant research.

Mount (127) discussed priorities for research in determining thermal requirements for freshwater fish. Desirable fish species in a particular environment were considered an economic crop. Priorities were suggested that would optimize production of harvested species and their supporting biota. Information on thermal requirements for these fish was deemed necessary at all stages of their development and for all functions required for maintenance of thriving populations.

Site Studies

Several papers described on-site field studies of biological effects resulting from thermal discharges.

Adams (2) summarized representative ecological studies carried out at thermal power stations discharging into estuarine and open ocean environments of California. He stressed the importance of undertaking field studies rather than relying entirely on laboratory data that had tenuous correlations with actual thermal discharges. Adams pointed out that if all 16,972 *Mw* of tidewater thermal generating capacity in California were operated at maximum capacity, a total of only 5.86 sq miles (15.3 sq km) of water surface area would be included within the + 2°F (– 16°C) isotherm.

Effects of heated discharges on freshwater fish in Britain were summarized by Alabaster (4). Laboratory and field studies were conducted at several thermal power stations. Potentially lethal temperatures at outfalls did not generally kill fish. Rather, fish either avoided temperature extremes or became acclimated to them. Fish were attracted to the discharges in cooler seasons. Indirect effects of heat were considered important, especially in already polluted waters.

Battelle-Northwest (15) published its annual progress report for 1968 on field and laboratory studies of the biological effects of thermal discharges into the Columbia River. Condensed reports were given on (*a*) responses of salmonid fish to acute thermal shock (C. C. Coutant, J. M. Dean), (*b*) passage of downstream migrating salmonids through Hanford thermal plumes (C. C. Coutant, C. D. Becker, E. F. Prentice), (*c*) spawning of chinook salmon near thermal discharges (D. G. Watson), (*d*) survival and growth of eggs and young chinook salmon under altered seasonal temperature regimes (P. A. Alson, R. E. Nakatani), (*e*) tracking of sonic-tagged salmon and trout migrating past thermal discharges (C. C. Coutant), (*f*) food of juvenile chinook salmon at Hanford (C. D. Becker), (*g*) columnaris exposure and antibody development among upstream migrant salmonids (M. P. Fujihara), (*h*) oral immunization of juvenile coho salmon against *Chondrococcus columnaris* (M. P. Fujihara), and (*i*) alterations in hepatocyte function of thermally acclimated rainbow trout (J. M. Dean, J. D. Berlin).

Beer and Pipes (17) conducted a brief survey of a thermal discharge from the Riverside generating station into Pool 15 of the Mississippi River near Davenport, Iowa. Effects of heat on the biology and chemistry of the pool seemed minimal, based on observations of heat dissipation patterns, water chemistry, bottom fauna, and fish. Recommendations were made for effluent discharges from a proposed generating station nearby.

Churchill and Wojtalik (39) discussed the effects of heated discharges from several thermal plants operated by the Tennessee Valley Authority. Studies included physical dispersion of heated water, plankton, periphyton, bottom fauna, and fish. Ecological considerations have played a large role since the late 1950's in planning new thermal discharges in the Tennessee Basin.

Nakatani (132) discussed investigations conducted on the Columbia River at Hanford, Wash., to determine the effects of heated discharges on anadromous fish. Hanford is the site of plutonium production reactors of the U. S. Atomic Energy Commission that have discharged heated water to the river since 1944. Nakatani concluded that data on juvenile and adult salmonids showed no evidence of thermal damage to the fishery resource as a whole. In a discussion of Nakatani's paper, Snyder (169) described applications of thermal pollution studies at Hanford to proposed thermal discharges elsewhere in the Columbia River. Hydroelectric and thermal power development plans were reviewed along with fishery problems that could be anticipated from flow and temperature alterations. Some research facilities of the Bureau of Commercial Fisheries were described.

Poltoracka (147) found that warming a Polish lake 5.9° to 9.6°C above normal seasonal temperatures by discharges from a thermal power plant increased the number of taxa of algae, especially Chlorophyta, compared to unheated lakes nearby. The largest number of species was found in the warmest lake, and the lowest in the coolest. Although the Chlorophyta fluctuated in percent composition throughout the 1-yr study, this algal group was consistently most abundant in the warmed lake.

Proffitt (150) reported a field investigation conducted on the White River, Indiana, at the site of a 220 *Mw* power station that used direct cooling (ΔT 10° to 11°C). Two yr of preoperational surveys preceded nearly 2 yr of postoperational studies. No detrimental changes occurred in DO, turbidity, BOD, macroinvertebrate, or fish fauna following plant startup although some changes were observed.

Moyer and Raney (128) summarized engineering and ecological considerations in designing thermal discharges from the Peach Bottom Nuclear Power Station, Units 2 and 3, on the Susquehanna River, Pennsylvania. They concluded that heated effluents [13°F (– 10.5°C) above ambient] could be introduced and water temperatures thus altered without causing serious ecological problems, provided necessary data were available and proper studies were completed, including ecological surveys and plume dispersion modeling.

Biogeography

The present world distribution of the brook trout, *Salvelinus fontinalis,* was reviewed by MacCrimmon and Campbell (112), who found that water temperature seemed to be the most important single factor limiting the distribution. Adequate precipitation and suitable spawning areas also were deemed necessary for establishing wild populations.

Unsuccessful attempts to plant trout (*Salmo irideus*) in ponds where summer temperatures exceeded 25°C were reported by Altukhov *et al.* (6). In contrast, the coregonid, *Coregonus peled,* thrived in the same ponds. Calderon (35) successfully cultured brown (*Salmo trutta*) and rainbow (*S. gairdneri*) trout at temperatures above 22°C. High oxygen requirements of the trout at these temperatures were met by supersaturating the water supply. Characteristics of incubation temperatures, rates of development, sizes vs. age, and management techniques were discussed.

Tolerance limits for temperature, salinity, DO, and current strength were found by Meldrim (122) not to restrict distribution of the Olympic mudminnow (*Novumbra hubbsi*) directly. Behavioral preferences for current and salinity seemed to be major factors, along with selection of light intensities, substrate type, and vegetative habitats.

Regan (154) studied the relationships of temperature, salinity, and other factors to the distribution of four species of euphausids from coastal waters of British Columbia. *Euphausia pacifica* was most tolerant of fluctuating environmental conditions in the field, followed in order of decreasing tolerance by *Thysanoessa spinifera, T. longipes*, and *T. raschii*. In the laboratory, specimens migrated vertically in increasing temperature gradients and entered higher temperatures than those encountered in the field.

Heubach (75) observed rapid summer declines in mysids, *Neomysis awatschensis*, inhabiting the Sacramento-San Joaquin estuary, California, when water temperatures exceeded 22°C. The decline apparently was not due to decreased reproduction since the percentage of gravid males remained high in each of the 2 yr of the study. Maximum population increases occurred between 15° and 20°C in the spring.

According to Winterbourn (198), the fauna of thermal waters of New Zealand were similar in composition to those reported from North America, the Himalayas, and Algeria. Petrovska (146) characterized 11 species of Cyanophyceae and two species of *Beggiatoa* growing together in thermal springs of Uranjska Banja. *Phormidium laminosum* dominated the microflora.

Dow (51) demonstrated correlations between cyclic sea temperatures from 1905 to the present at Boothbay Harbor, Maine, and the commercial catches of the American lobster along the northeast Atlantic coast. Rising temperatures from 1940 to 1953 increased lobster yields in the northern part of the range while yields to the south declined. The recent 16-yr cooling trend has reduced northern catches and increased southern ones. Although the entire range of the species may have been unaffected by cyclic changes, population sizes apparently were strongly influenced. Mukhin (129) related the size of commercial catches of cod and haddock in the Barents Sea to fish population sizes, and to water temperatures that determined fish location relative to fishermen.

In a field survey of southwest British Columbia, Ellis and Borden (56) found the freshwater hemipteran, *Notonecta undulata*, to live and reproduce under a wide variety of ecological conditions. First and fifth instars and adults preferred 27° ± 1°C. Adults survived 34°C for up to 9 hr, while resistance times above 34°C decreased with increased temperature.

Kenny (91) found that high environmental temperatures, even at low tide, did not seem to restrict distribution of the esuarine polychaete, *Clymenella torquata*, directly. The temperature at which 50 percent survived a 5 min exposure was 39.2°C, whereas it was greater than 35°C for a 2 hr exposure. Substrate surface temperatures in the field reached 36.9°C but were only 33.4°C, well below the lethal level, at the 10-cm depth of most worms. Tolerance to low salinity increased at low temperatures characteristic of winter.

Synergism

Environmental temperatures affected the response of bluegills, *Lepomis macrochirus*, to gradual oxygen depletion in studies by Spitzer *et al.* (172). Respiration of fish acclimated to 13° and 25°C showed varying degrees of oxygen independence while those at 30°C were oxygen dependent. Ventilation rate increased in 13° and 25°C fish but not at 30°C. Heartbeat showed initial increase in 10° and 25°C fish, but immediately decreased in 30°C fish. Fish adapted to cooler water exhibited response thresholds for each rate at lower oxygen tensions.

The susceptibility of bluegills and rainbow trout to 12 pesticides tested by Macek *et al.* (113) generally increased during the first 24 hr as temperatures increased from 12.7° to 23.8°C and 1.6° to 12.7°C, respectively. However, the susceptibility of bluegills to lindane and azinphosmethyl was unaffected by temperature and susceptibility of both species to methoxychlor decreased as temperature increased. This general temperature effect was usually less after 96 hr of exposure, with the exception of susceptibility of rainbows to trifluralin.

Hussein *et al.* (78) found that lower water temperatures delayed and diminished the toxicity of toxaphene to the fish, *Gambusia* sp. and *Tilapia zilli*. At 32°C, 0.072 mg/l caused 100 percent mortality but at 16°C, 0.165 mg/l was required.

Recovery of the fish *Oryzias latipes* from exposure to 3 or 4 *k* rad of X-rays seemed to be related to replacement of damaged cells in critical organs, a process shown by Egami (55) to have a rate dependent on temperature. When amoebae, irradiated with UV at 4,800 ergs/sq mm, were cooled at 6°C for 2 hr by Skoeb (165), they had better survival and higher concentrations of RNA and protein than nonchilled amoebae. A restorative effect of cold was claimed.

Shabalina (160) demonstrated that chronic oral administration of cobalt (as chloride) affected thermal resistance of trout (*Salmo iridens*). Doses of 0.0044 mg/day/kg body weight lowered the undefined "critical temperature" of first-year trout by 2.5°C. Doses of 0.10 mg/day/kg raised that level by 1.6°C. There were corresponding changes in fat content, blood hemoglobin, red cell counts, and iodine number.

Harvey (69) demonstrated that nonlethal variations in water temperature (25°, 30°, 35°, 40°C) had no major influence on sorption of the radioisotopes ^{137}Cs, ^{85}Sr, ^{65}Zn, ^{59}Fe, ^{57}Co, and ^{54}Mn by the blue-green alga, *Plectonema borganum*. With increasing temperature, there was some decrease in uptake of ^{57}Co and an increase of ^{54}Mn (except at 40°C). There were no changes in growth rates of cultures attributable to radioactivity.

Increased temperatures increased the sensitivity of brine shrimp, *Artemia salina*, to aflatoxin according to Brown *et al.* (29). Optimum sensitivity occurred at 37.5°C.

The combined effects of salinity and temperature on survival and growth of mussel (*Mytilus edulis*) larvae were studied by Brenko and Calabrese (23). Survival of larvae at salinites of 15 to 40 ppt was good (70 percent or better) between 5° and 20°C, but mortalities increased at 25°C. High or low salinities aggravated lethal effects at high temperatures. Optimum larval growth occurred at 20°C in salinities from 25 to 30 ppt. Growth decreased at 10° and 25°C, especially at high (40 ppt) or low (20 ppt) salinities.

Resistance

The highest maximum thermal tolerance ("critical thermal maximum") recorded for a fish, (44.6 ± 0.05°C) was determined by Lowe and Heath (109) for *Cyprinodon macularius* from the Sonoran Desert, Ariz. The natural daily thermal cycle produced higher tolerances than did constant-temperature laboratory acclimation. Thermal environments in the field that were within 2° to 3°C of thermal death were commonly selected.

Bidgood and Berst (19) identified no difference in tolerance to upper lethal temperatures among progeny from wild rainbow trout (*Salmo gairdneri*) that homed to four widely separate watersheds in the Great Lakes in the U. S. All eggs were incubated and juveniles reared under similar conditions and acclimated to 15°C. Physiological observations agreed with previously determined similarities in phenotypic characteristics among the four watersheds.

Westin (196) investigated the thermal tolerance (14 hr TL_M) of Baltic Sea fourhorn sculpins, *Myoxocephalus quadricornis*, and found that fish acclimated from 2° to 20°C showed a linear increase in lethal temperature from 17.5° to 25.5°C. Acclimation above 20°C produced no increased tolerance. The lower lethal temperature was above 0°C when fish were acclimated above 22°C. The thermal polygon for this species indicated a relatively small thermal tolerance zone befitting an organism obtained from a cold, stable environment.

Acute thermal shock of 7.5°C (from 27° to 34.5°C) to early embryonic stages of zebra fish for 2.5 hr caused increased mortality and abnormal embryos in experiments by Ingalls *et al.* (80). The highest mortality (78.7 percent) and abnormality (50.8 percent of hatch) of embryos occurred when heat treatment was begun at the 4- to 8-cell stage. Many types of deformities appeared, including scoliosis, crooked tail, rumplessness, short tail, and conjoined twinning. Observations were related to mammalian birth defects.

Gorodiliv (63) investigated the sensitivities of different embryonic stages of *Salmo salar* and *Coregonus peled* to thermal shock. A sensitive period began after midblastula and continued to the "eyed" stage. Maximum sensitivity occurred during full epiboly. There was a long lag before manifestation of effects incurred between fertilization and blastula, while death followed immediately after injury at later stages. Tatarko (182) studied the sensitivity of pond carp (*Cyprinus carpio*) to elevated temperature at different stages of embryonic development. Fertilized eggs were incubated at 20°C and exposed to 30° to 32.5°C. Two periods of special sensitivity were observed, one from the beginning of cell division to gastrulation and one from formation of the tailbud to hatching. High immediate mortalities and percentages of abnormal fry resulted.

Eggs of the sturgeon, *Acipenser guldenstadti colchicus,* were subjected to various exposures of heat shock at 34°C by Vasetskii (191), and survival and ploidy were determined. All larvae hatching from eggs heat-treated as oocytes had 2 *n*. chromosomes. Oocytes treated for more than 40 min did not mature into viable eggs. Haploid, diploid, and triploid complements of chromosomes were observed in larvae from eggs heated just prior to fertilization. The triploid frequency was 15 to 20 percent. Survival of mature eggs decreased with duration of exposure, and all eggs were killed within 11 min.

According to Stephenson and Potter (174), the upper temperature limit for maintenance *in vitro* of heart, liver, and kidney tissues from the ammocoete of the lamprey, *Mordicia mordax,* was related to the highest temperatures encountered in the field by the larvae. Optimum heartbeat and cell growth occurred at 25°C, while 30°C was marginal for tissue viability, and 37°C was lethal within 12 hr.

Johansen (84) investigated the role of the pituitary in thermal resistance of the goldfish. Fish with intact pituitaries displayed acclimation following repeated exposures to higher temperature. Hypophysectomy before the exposures prevented acclimation in otherwise similar test fish.

Haefner (66) determined temperature and salinity tolerances of the sand shrimp, *Crangon septemspinosa.* Male and female shrimp were subjected to 12 temperature-salinity combinations within the ranges of 4° to 23°C and 5 to 40 ppt. Mortality data were fitted to response surfaces to estimate the combinations providing maximum survival. Both sexes tolerated wide ranges of salinity (19 to 36 ppt, 18 to 39 ppt, males and females, respectively) between 10° and 14°C, but lower or higher temperatures reduced this range. The laboratory data compared favorably with known aspects of the ecology and distribution of the species.

Tagatz (179) determined the 48-hr median upper and lower thermal tolerance limits of adult and juvenile blue crabs (*Callinectes sapidus* in relation to extremes of salinity and acclimation temperature. Low salinity (20 percent seawater) reduced tolerance to high and low temperatures by 0.2° to 2.0°C, compared to undiluted seawater. Crabs displayed thermal acclimation at 6°, 14°, 20°, and 30°C, and their thermal polygon exceeded in area those of most fishes. Tolerance limits for juveniles and adults were similar.

Reed (153) defined optimum ranges of temperature and salinity for laboratory-cultured zoeae of the Dungeness crab (*Cancer magister*) at 10.0° to 13.9°C and 25 to 30 ppt, respectively. Zoeal survival was not affected significantly by temperatures and salinities approximating normal ocean ranges off the Oregon coast during the larval period. Growth rates of crab larvae were related directly to temperature.

The influence of laboratory acclimation to various temperature-salinity combinations (22° or 35°C, 5 or 30 ppt) on heat tolerance of the hermit crab, *Diogenes bicristimanus*, was determined by Nagabhushanam and Sarojini (131). A combination of high temperature and high salinity was most favorable for thermal resistance. Either lower salinity or lower temperature during acclimation reduced resistance.

McRitchie (121) demonstrated increases in the upper lethal temperatures of the estuarine gastropod, *Thais haemastoma,* when the acclimation temperature was raised from 14° to 23° and 34°C.

Thermostability of muscles of *Mytilus edulis* and *Macoma baltica* increased 50 and 63 percent, respectively, in studies by Makhlin and Skholl (116), after temperature elevation of 7° to 11°C for four days. A decrease to previous levels occurred after 8 to 10 days. Thermostability was defined as the time to loss of excitability at a lethal temperature, 39°C for *M. edulis* and 37°C for *M. baltica.* Thermostability of *M. baltica* muscle decreased 43 percent following 7°C temperature reduction but returned to original levels after 8 days. There was no change in aldolase activity in muscle following temperature increases.

The survival of oysters buried in 7.6 cm of marine soil at five temperature ranges was determined by Dunnington (52). Accidental burial of oysters is a common, natural, and man-caused phenomenon. Survival time varied from 2 days in summer (over 25°C) to 5 wk in winter (less than 5°C), and there were proportional changes in decomposition rates.

Gonor (62) reported studies of temperature relations of various Oregon marine intertidal invertebrates. Principal emphasis was directed to microenvironmental temperatures in the natural habitats of limpets, chitons, mussels, barnacles, snails, and sea urchins, and the internal temperatures of these animals. All species sustained temperatures as much as 10°C above sea surface during low tides in early summer.

Temperature resistance, capacity adaptation, and diapause were investigated in aquatic and terrestrial insects and snails by Precht (149). Little uniformity of response to temperature was found among the many species.

Seasonal and geographic variations in resistance to high temperatures of two species of intertidal, tube-dwelling polychaetes were identified by Kenny (92). Median survival temperatures in winter at Beaufort, N. C., were about 4°C less than the summer survival temperatures of 42.5°C (*Diopatra cuprea*) and 40.5° (*Clymenella torquata*). Worms from Massachusetts and North Carolina revealed more than 4°C difference in summer survival temperatures. Experimental acclimation in the laboratory duplicated field acclimation.

The influence of different salinities and temperatures on spermatozoa of *Nereis rubens* and *Styela rustica* was studied by Khlebovich and Lukanin (93). Viability in the optimal salinity range (8 to 24 ppt) in all cases decreased as temperature increased between 11° and 20°C. Decrease in viability probably was due to enhanced energy expenditure.

Pattee (141) found the upper incipient lethal temperature for the freshwater planarian, *Dugesia gonocephala*, to lie near 20° or 21°C in stirred, aerated water. This level of thermal resistance was intermediate between the planarians, *Polycelis felini* and *P. nigra*, which correlated with field distributions.

Many stocks of *Paramecium aurelia* were analyzed for differential thermal resistance by Lavatelli *et al.* (102). Clear differences, believed to be hereditary, were obtained at 36° and 38°C. Crosses then were conducted with temperature-resistant strains by Crippa (44). Analysis of F_1 and F_2 generations indicated a dominant involvement of cytoplasm in thermal resistance, for the ex-conjugant lines derived cytoplasmically from a resistant parent retained the acquired character for 60 fissions after conjugation (non-autogamous lines). Some ex-autogamous lines kept the acquired character for more than 30 fissions after autogamy.

Sopina (171) recognized different thermal resistance times at 41°C for several clones of *Amoeba proteus* cultivated at 17°C. The characteristic resistance time of each clone persisted through prolonged cultivation and subcloning. The fact that nuclear transplantation did not alter the relationships suggested cytoplasmic inheritance.

Bai *et al.* (11) increased the temperature of cultures of the ciliate protozoan, *Blepharisma intermedium*, 1°C/day and observed activity, structure, reproductive rate, and cytochemistry. The reproductive rate increased from 24°C, the initial temperature, to an optimum at 28°C, whereafter the rate declined to zero at 38°C. Requisite quantities of glycogen, basic proteins, unsaturated lipids, and alkaline and acid phosphatases were present up to 38°C, but cytochemistry and cell morphology were altered above that temperature. Size decreased progressively above 30°C, and activity was curtailed above 33°C. Culture organisms recovered from exposure to 38°C on return to room temperature although there was a 3-day lag period.

Freshwater protozoan communities were exposed to temperature shocks by Cairns (33) who determined the time for return to preshock levels of species diversity. The magnitude of shock was more important in reducing diversity than duration of exposure. Restoration of diversity required only a few hours when the shock was mild (elevation from 20°C to 30°C) but 5 or 6 days for severe shocks (to 50°C).

Thermal acclimation of heat and cold resistance in some marine algae was demonstrated by Lyutova *et al.* (111). Heat resistance, but not cold resistance, varied with incubation temperatures of the sublittoral *Laminaria saccharina* and *Chorda filum* and the deep water form of *Fucus vesiculosus*. Both heat and cold resistance changed in *Ascophyllum nodosum*. Cold resistance of *L. saccharina* changed seasonally in nature, increasing with decreasing environmental temperatures.

Levina (105) evaluated the survival of human bacteria in river water at various temperatures. *Salmonella typhosa* survived 18 days at 10°C while *E. coli* survived for 22 days. *Streptococcus faecalis* survived 27.6 days at 1°C but only 14 days at 20°C.

Parameters of thermal death were determined by van Uden *et al.* (190) for 10 strains of yeast. Maximum temperatures for growth (Tmax) ranged from 22° to 49°C, and there were high correlations in all but one case between specific thermal death rate and the respective values. A new parameter, the "thermal death activation constant," was introduced to characterize the relationship.

Reproduction

Vincent (193) studied the spawning ecology of white bass, *Roccus chrysops*, in Utah Lake, Utah. Schooling of males first was observed when water temperatures reached 11.1°C in April, but became pronounced only at 13.3°C in late April. The first gravid females appeared over spawning beds in early May when temperatures reached 17.2°C. Spawning areas were homothermic during initiation of activity which continued until mid-June at 20.6°C.

According to Hyder (79) temperature and day length were the most important factors influencing gonadal development and reproductive activity in the cichlid fish, *Tilapia leucosticta*, in an equatorial lake. Berenbein (18) described a close relation, characterized by a high coefficient of correlation, between the time of spawning of some herring subspecies in the Atlantic and Pacific Oceans, and values for the mean temperature of the seawater surface in the spawning areas.

Billard (21) determined the relationships of temperature to the duration and efficacy of spermatogenesis in the guppy, *Poecilia reticulata*. Within the limits of 20° to 30°C, spermatogenesis was accelerated by temperature increase. The temperature optimum for sperm yield was 25°C.

The influence of temperature on gonad development in the three-spined stickleback (*Gasterosteus aculeatus*) was found by Schneider (159) to be modified by photoperiod. High temperature stimulated gametogenesis in both sexes when tested in combination with a long photoperiod, but gametogenesis ceased with a short photoperiod. Low temperature led to slight gonadal development regardless of photoperiod.

Survival of pike (*Esox lucius*) eggs incubated in various thermal regimes was determined by Lillelund (107). The number of degree-days at a constant temperature required to attain any given embryonic stage increased exponentially with decreasing temperature. High survival, not influenced by temperature, occurred between 9° and 15°C. There was 40 percent mortality at 5.8°C, 7 to 29 percent mortality at 18°C, and 66 to 81 percent mortality at 21°C. Diurnal oscillations between 15° and 20°C resulted in 12 percent mortality.

Experiments by Reznichenko *et al.* (155) defined the upper and lower limits for successful constant temperature incubation of tench (*Tinca tinca*) eggs. The lower limit seemed to lie between 15° and 18°C, while the upper limit fell between 30° and 35°C. Best survival occurred at temperatures most common in the natural environment during incubation.

Aiken (3) determined that temperature and photoperiod controlled ovarian growth and maturation in the crayfish, *Orconectes virilis*, from Alberta, Canada, and that increased water temperature induced egg laying in the spring. Complete ovarian maturation required 4 to 5 months of low temperature (4°C) and constant darkness. Ovarian growth occurred in warm water and darkness, and in cold water and long-day photoperiod, but subsequent exposure to spring temperature and photoperiod would not then induce egg laying.

Schaller (158) found that the duration of embryonic diapause in the dragonfly, *Aeschna mixta*, depended on the temperature condition to which the eggs were subject after oviposition. A temperature drop to 5° to 10°C allowed continuation of development during diapause, which then completed rapidly after return to 20°C. Development at a constant 20°C proceeded much more slowly. About 5 wk were required at low temperature to shorten embryonic development. Prolonged exposure to cold resulted in synchronization of egg hatching after return to 20°C. Anderson (7) discovered that short day length, not temperature, induced embryonic diapause in the mosquito, *Aedes atropalpus*.

Changes in gonad size and initiation of gametogenesis, recorded by Webber and Giese (194) in the black abalone, *Haliotis cracheroidii*, from Monterey County, Calif., showed no apparent correlation with seasonal water temperature. There was also no correlation between gonad index and water temperature data for 10 yr of study of the chiton, *Katharina tunicata*.

Moore and Reish (125) studied seasonal variations in gamete maturity in mussels, *Mytilus edulis*, from Alamitos Bay, Calif. Mature ova were present only from November through May. Mature sperm were found in some mussels throughout the year, but the greatest development occurred between October and February. The higher the temperature the greater the number of sexually indeterminate individuals.

Distinct races of oyster, *Crassostrea virginica*, along the eastern coast of the U. S. required different temperature regimes for completion of gametogenesis and spawning, according to Loosanoff (108). Some Long Island Sound oysters ripened at 12°C, but New Jersey and more southern groups (North Carolina and Florida) did not. At 15°C, 60 percent of the Long Island oysters spawned after 45 days while the New Jersey oysters, even after 72 days, produced only 20 percent with recognizable gonads, and the more southern groups were even less developed. Differences were also shown at 18°C.

Polymorphism

Changes with temperature in mean meristic characters of plaice (*Pleuronectes platessa*) were identified in the laboratory by Molander and Swedmark (124). Both high and low incubation temperatures produced higher counts of vertebrae. The number of rays in the anal and dorsal fins increased with rising temperature while the number of complex vertebrae decreased. Morphological variations in isolated plaice populations were explained by these observations in preference to recognizing genetic races.

Harrington (68) defined the thermolabile period for sex determination and differentiation in the ontogeny of the normally hermaphroditic fish, *Rivulus marmoratus*. Eggs incubated at a constant temperature of 20°C or higher yielded only hermaphrodites while lower temperatures yielded some males. Selected temperature changes during ontogeny established the thermolabile period between stage 31 (neural and hemal arches on caudal vertebrae) and hatching.

The relationship of temperature (18° to 30°C) to incidence of bisexuality in the heterogenic rotifer, *Asplanchna sieboldi*, was investigated by Buchner and Kiechle (30). Bisexuality varied between 3 and 30 percent, with the periodic cycle faster and fluctuations somewhat wider at warmer temperatures.

Toriumi (185) found that the number of spines of newly-formed spinoblasts in the ectoproct, *Lophopodella carteri*, increased following exposure of colonies to solar warming. Conversely, cooling the colonies decreased spine number. These aftereffects of temperature change occurred over the normal range of environmental temperatures.

Lewis and Jenkins (106) studied temperature-controlled polymorphism in the foraminiferan, *Nonionella flemingi*. Adults from the continental shelf of northern New Zealand were small and trochospiral while those from subantarctic islands were commonly twice as large and planispiral. Mean length and number of chambers in the final whorl of a local population was related inversely to environmental temperature. Reproductive maturity was reached at earlier stages in lower, colder latitudes.

Abbott (1) discovered that increasing and decreasing the temperature from acclimation caused aggregation of melanin in melanophores of isolated scales from the fish, *Fundulus heteroclitus*. Light had no effect. The reaction, which lightened the color of the skin, was reversible. The isolated melanophores apparently acted as independent effectors.

Tatarko (181) showed that meristic characters of carp, *Cyprinus carpio*, were altered by temperature conditions during development. The effects were manifested in number of vertebrae; lateral line scales; rays in dorsal, anal, pectoral and pelvic fins; pharyngeal teeth, and gill rakers. Number changed with both high and low temperatures during incubation, but there was no general correlation of number and temperature. Caution was recommended in using thermolabile morphological features in taxonomy.

Feeding

Rates of filtration and food ingestion were determined by Winter (197) for the lamellibranchs, *Arctica islandica* and *Modiolus modiolus*, in relation to temperature, concentration of particles, and body size. Temperature coefficients for filtration rate were 2.05 (at 4° to 14°C) and 1.23 (at 10° to 20°C) in *A. islandica*, and 2.33 and 1.63, respectively, in *M. modiolus*. For ingestion of algae, coefficients were 2.15 (at 4° to 14°C) and 1.55 (at 10° to 20°C) for *A. islandica*, and 2.54 and 1.92, respectively, for *M. modiolus*. A 50 percent reduction in both parameters occurred with a temperature drop from 12° to 4°C. The algae concentration allowing maximum rates of filtration and food utilization increased with temperature, but there was a threshold particle concentration for optimal use at each temperature.

Mangum (118) reported low temperature blockage of feeding response in two species of nereid polychaete worms at temperatures well within the zones of thermal tolerance. Blockage occurred between 10° and 15°C in *Nereis succinea* from the York River, Va., and near 0.°C in *N. vexillosa* from Kitoi Bay, Alaska. This behavioral character may limit the geographic range of a species.

Filtering rates of four species of *Daphnia* were measured by Burns (32) at 15°, 20°, and 25°C. Maximum filtering rates increased with warming temperatures and larger body sizes in all species. When rates were expressed on a unit-body-weight basis, differences were revealed among species. Filtering rates of adult *D. schødleri* and *D. pulex* were similar, and at 20°C were slightly higher than at 15° and 25°C. In contrast, filtering rates of *D. magna* and *D. galeata mendotae* increased with increasing temperature (Q_{10}'s of 2.38 and 2.71, respectively).

Shrable *et al.* (163) found the optimum temperature range for rapid digestion of dry pellet food by channel catfish, *Ictalurus punctatus*, was 26.6° to 29.4°C. The rate was slightly less at 21.1° and 23.9°C but considerably less at 10.0° and 15.5°C.

Growth

The growth of young sockeye salmon (*Oncorhynchus nerka*) was studied by Brett *et al.* (24) at temperatures ranging from 1° to 24°C in relation to food rations of 0, 1.5, 3, 4.5, and 6 percent/day of dry body weight, and at an "excess" ration. Optimum growth occurred near 15°C for the two highest rations, and shifted progressively to a lower temperature at each lower ration. The maintenance ration increased rapidly above 12°C, and amounted to 2.6 percent/day at 20°C. No growth occurred at 23°C despite availability of excess food. Maximum net food conversion efficiency of 40 percent occurred at 8° to 10°C for rations of 0.8 to 1.5 percent/day.

The importance of feeding relationships and temperature optima for growth of fish were stressed by Doudoroff (50) in a discussion of thermal requirements of fish. Examples were given of changes in optimum temperatures for growth when the food ration was altered. Juvenile coho salmon grew best at 17° to 20°C on an unrestricted ration, while best growth occurred at the lowest temperature range tested, 5° to 8°C, at reduced rations.

Ouchi (136) found that there was no scale growth in goldfish below 12.5°C. Growth rate of the entire scale increased at progressively higher temperatures between 12.5° and 27.5°C, while the time required to form one scale ridge decreased. Ridge spacing was unrelated to water temperature. The results were applied to use of scales for determining age and life history of fish.

Savitz (157) determined the effects of temperature and body weight on maintenance protein requirements by bluegill (*Lepomis macrochirus*) acclimated to 7.2°, 15.6°, 23.9°, and 29.4° to 32.3°C. Endogenous nitrogen excretion rates (in nitrogen-free diet) generally decreased at progressively lower temperatures, but remained constant between 15.6° and 7.2°C. Seasonal changes in protein requirements for a natural bluegill population were calculated.

The effect of temperature on preadult development of the mosquito, *Aedes flavescens,* was investigated by Trpis and Shemanchuk (187). The most favorable temperature was 20°C, at which 74 percent of the larvae pupated in 20 days. Development periods and survival were 80 days and 6 percent at 10°C, 42 days and 54 percent at 15°C, and 22 days and 7 percent at 25°C. At 5° and 20°C, mortality was 100 percent in the first instar. Trpis and Horsfall (186) found that development of mosquito, *Aedes sticticus,* proceeded with least mortality when temperatures fluctuated about a mean of 20°C. Larvae at instars 1 and 2 withstood cooler water than those in later instars. This pattern correlated with maturation at rising temperatures in nature.

The effect of temperature on growth, longevity, and egg production in freshwater anostracan, *Chirocephalus diaphanus,* was studied in the laboratory by Lake (101). Maximum growth and earliest sexual maturity occurred at the highest temperature tested, 25°C. The greatest longevity, greatest final length, and highest egg production occurred at 10°C. Few eggs were produced at 5°C. The most favorable range was considered to be 10° to 20°C.

Fagetti (58) found that a 5°C elevation of seawater temperatures prevented completion of larval development of the spider crab, *Pisoides edwardsi.* The highest mortality occurred in the second zoeal stage, but higher temperature regimes speeded development prior to mortality.

The development of larval stages of penaeid shrimp was affected by temperature according to Cook and Murphy (42). At 3 ppt salinity, the average time required in the laboratory for larvae of brown shrimp, *Penaeus a. aztecus,* to reach the first postlarval stage was 17 days at 24°C, 12.5 days at 28°C, and 11 days at 32°C. Survival of nauplii was maximum at 24°C. As the larval shrimp advanced to protozoeae and myses, survival rates usually increased with a rise in temperature.

Davis and Calabrese (46) evaluated the survival and growth of larvae of the European oyster (*Ostrea edulis*) at different temperatures. There was satisfactory survival (>70 percent) at 12.5° to 27.5°C, whereas satisfactory growth (70 percent or more of optimum) was attained at 17.5° to 30°C. The greatest number of attached spat were obtained at 20° to 22.5°C. Spat kept at 10°C showed virtually no growth, but growth increased with each increase in temperature between 12.5° and 27.5°C. Approximate spat setting times were 26 days at 17.5°C, 14 days at 20°C, and 8 to 12 days for 25°, 27.5°, and 30°C.

Temperature was one of the environmental and biological controls of bivalve shell minerology considered in a review by Kennedy *et al.* (90). Generally an inverse relationship occurred between the percentage of calcite in the shell and the mean temperature of the environment. This relationship was evident for a wide variety of species, both in total shell structure and in layers secreted during different seasons.

High temperatures were observed by Biernacka (20) to stunt growth and shorten development time of a variety of protists in the Baltic Sea and in the Vistula lagoon, but the effect was not uniform among all species. Salinity changes seemed to have a greater effect on distribution than did temperature.

Kalyuzhnaya *et al.* (87) studied the effects of temperature on growth and antibiotic formation in 87 strains (65 species) of the fungal group Actinomycetes found in soil and water. Optimal temperatures for development between 28° and 37°C, and for formation of antibiotics between 18° and 28°C, was demonstrated. Optima among species, however, revealed considerable variation, and there was essentially no growth below 6°C.

Growth of bacteria living above 90°C in boiling springs at Yellowstone National Park was reported by Bott and Brock (22). Cells were grown on microscope slides immersed in the springs, and ultraviolet radiation was administered periodically to aid in distinguishing between growth and passive attachment. Estimated generation times ranged from 2 to 7 hr, values that were comparable to those of aquatic bacteria living in cooler environments.

Broch and Freeze (27) described a new species of thermophilic, aquatic bacterium, *Thermus aquaticus*. The species was isolated from thermal springs at Yellowstone National Park, from springs in California, and from man-made thermal habitats such as hot tap water. Optimum temperature for growth was 70°C, the maximum 79°C, and the minimum about 40°C. Generation time at optimum bacterial growth was about 50 min. Maheshwari (115) isolated six species and three varieties of thermophilic fungi from India, and reported their temperature ranges for growth.

Physiological properties and temperature responses of the bacterial epiphyte, *Leucothrix mucor*, isolated from a tropical seacoast were reported by Kelly and Brock (89). The minimum temperature for growth was higher in the tropical (13°C) than cool water (2° to 5°C) strains. The stenothermal nature of the tropical strain was related directly to the equable temperature regime (25° to 30°C) in its environment, compared to the eurythermal nature and broader environmental temperature range of cool-water strains.

Growth rates of bacteria isolated from natural water, and their relationships to temperature, were investigated by Baig and Hopton (12). A "temperature characteristic", μ, was derived by substituting growth rate for reaction rate in the Arrhenius equation, where μ corresponds to the activation energy of a chemical reaction. Organisms with fastest growth at the low temperature of from 5° to 10°C had the lowest temperature characteristics, while the correlation between growth and μ was best over the range of 15° to 25°C. Psychrophily thus was quantified in the low temperature characteristic.

Temperature Selection

Local migrations of the teleost, *Tilapia grahami*, in response to diurnal temperature changes in shallow water were observed in Lake Magadi, Kenya, by Coe (41). These thermally adapted fish preferred temperatures of 36° to 43°C and exhibited feeding to a browse line that closely coincided with upper lethal limits. Fish migrated out of shallows when night temperatures dropped gradually to near 23°C and entered zones of inflowing thermal water at 43°C.

Leggett and Power (103) detected seasonal movement of landlocked Atlantic salmon into deep waters of two Newfoundland lakes when the surface warmed above 14°C. In one lake, benthic dispersal contributed to an early end to the growing season.

Pink salmon (*Oncorhynchus gorbuscha*) fry tested in vertical temperature and salinity gradients by Hurley and Woodall (77) selected a restricted range of temperatures that shifted downward as fish size increased. The youngest fry generally selected 11.7° to 13.3°C, older fry 9.4° to 10.6°C. Increasing salinities were selected in sequence as fry made the transition from freshwater to seawater. Results were related to typical behavior and survival of pink salmon fry in the Fraser River, B. C.

Hammel *et al.* (67) concluded from experiments with the arctic sculpin, *Myoxocephalus*, that the temperature of a rostral part of the brain activates a response to escape from a warmer than optimum environment for this species. Without experimental alteration of forebrain temperature, the fish departed elevated temperatures (12°, 16°, or 20°C) at about 8°C. Rapid elevation of brain temperature independent of the body caused earlier departure, and lowering of only brain temperature induced prolonged stays at high environmental temperature.

Physiology

A number of papers dealt with temperature compensation in respiratory metabolism. Edwards *et al.* (54) found that underyearling plaice, *Pleuronectes platessa*, were cold-adapted over a 5° to 10°C range, with 20°C as a possible upper limit for survival. Respiratory rate (ml O_2/h) in the normal cold range was $0.214\ W^{0.721}$ (where W is live weight in grams). There was little adaptation of respiratory rate above 15°C, and high Q_{10}'s were exhibited.

Initial adaptive responses in oxygen consumption of live fish to warm and cold stress were studied by Parvatheswararao (139). *Etroplus maculatus* was studied in the summer, acclimated to 20° and 35°C. *Cirrhina reba* was studied in winter, but also acclimated to warm temperatures. Responses were discussed in relation to thermal history and the thermal gradients imposed in the experiments.

A variety of experiments concerned with metabolic temperature compensation in fishes was described by Jankowsky (81). Temperature acclimation was demonstrated in the golden orfe (*Idus idus*) for rates of oxygen uptake in excised gill and muscle tissue, activity of cytochrome oxidase in skeletal muscle, and activities of aldolase and malic dehydrogenase in gill, liver, and muscle homogenates. Metabolic rate of muscle *in vitro* did not completely reflect the capacity adaptation of the intact fish, as determined by examining frequencies of opercular and cardiac cycles, and local oxygen pressure in the dorsal skeletal muscle, heart cavity, and caudal blood vessels.

Haschemeyer (71) identified a metabolic compensation by the toadfish (*Opsanus tau*) to low temperature in experiments on oxygen consumption in a closed respiratory chamber. Cold-acclimated (10°C) toadfish were found to consume oxygen at a rate 42 percent greater than 21°C-acclimated fish when both were tested at 22°C.

Peterson and Anderson (145) determined oxygen consumption of brain tissue from Atlantic salmon (*Salmo salar*) and brook trout (*Salvelinus fontinalis*) at temperatures ranging from 6° to 30°C for fish acclimated to 6° and 18°C. Thermal acclimation provided nearly complete compensation for respiration of brain homogenate in sucrose, but only partial compensation of brain mince in saline. The difference was ascribed to an increase in ATPase activity in saline.

Peterson and Anderson (144) measured the effects of rapid change in temperature on the spontaneous activity and oxygen consumption of Atlantic salmon (*Salmo salar*) underyearlings acclimated to 6° and 18°C. The new temperatures ranged from 6° to 30°C. Spontaneous activity exhibited a transient rise during temperature change, followed by a generally lower stabilized phase which progressed gradually to a level characteristic of the new temperature. The peak of transient activity was correlated with the rate, rather than amount, of temperature change. At similar test temperatures, the standard metabolic rate (O_2 consumption) of fish acclimated at 6°C was higher than fish acclimated at 18°C. Complete acclimation for both parameters required about 2 wk regardless of the direction of temperature change.

Oxygen intake of the gills of the silver eel (*Anguilla vulgaris*) shifted noticeably after temperature changes as little as 4°C were administered to either the head region or the body, according to Leicht (104). A shift also occurred in the body skin (also oxygen absorptive) after an 8°C change in the body. Oxygen uptake appeared to be regulated actively through oxygen tensions of venous blood.

Job (83) investigated the effect of temperature (as well as of size, salinity, and partial pressure of oxygen) on the respiratory metabolism of the tropical fish, *Tilapia mossambica*. Oxygen uptake increased from 15° to 30°C, and became temperature-independent at 30° to 40°C as long as the pO_2 remained above 150 mm Hg. The thermal effects on respiratory metabolism were highly dependent on the different partial pressures of oxygen, a point not stressed by previous investigations.

The influence of temperature on energy and material utilization of young coho salmon (*Oncorhynchus kisutch*) was studied in the laboratory by Averett (9). Specific dynamic action was related to both temperature and consumption rate; it accounted for up to 17 percent of the energy of assimilated ration at high temperatures. Standard metabolism increased with temperature and decreased with increasing fish size. Consumption rate was correlated closely with energy expended in activity. Calculated energy budgets indicated seasonal changes in optimum efficiency of food utilization for growth when consumption was within temperature ranges similar to those in nature, e.g., 5° to 14°C in early spring, 11° to 14°C in early summer, 14° to 17°C in late summer, 11° to 17°C in fall, and 5° to 8°C in late winter. Efficient food utilization occurred at higher temperatures but required consumption rates that were unrealistically high for nature.

Grigg (65) identified a consistently higher oxygen affinity of blood from 24°C-acclimated brown bullheads, *Ictalurus nebulosus*, than from 9°C-acclimated ones, when tested at the same temperature. This shift, which seemed to be located at the erythrocyte and which accompanied thermal metabolic acclimation, compensated for the normal decrease in blood oxygen affinity with temperature rise.

Despite general consensus that insects do not adapt metabolically to temperature changes, Buffington (31) identified adaptive acclimation in larvae of the mosquito, *Culex pipiens pipiens*. Temperature compensation occurred outside the range of acclimation. Seasonal selection in natural populations was deemed responsible for further changing responses to temperature at the population level. Acclimation and compensation changed with larval age.

Temperature effects on oxygen uptake by three species of barnacles during their normal seasonal and physiological cycle was measured by Barnes and Barnes (14). The Q_{10} at 5°, 10°, 15°, and 20°C varied widely, even in a given species. *Balanus balanoides* and *Chthamalus stellatus* displayed marked *homeostasis* at certain seasons, while homeostasis was less evident in *B. balanus*. Variations in temperature responses by metabolism were related to seasonal changes in nutritional and reproductive states.

Pamatmat (137) identified seasonal adaptation in respiration of the intertidal, sandflat bivalve *Transennella tantilla*. Cold adaptation corresponded to Prosser's pattern III A, consisting of clockwise rotation of the respiratory rate vs. temperature line. There was an unusual intersection of the cold-adapted and warm-adapted rate-temperature lines within the normal range of habitat temperature. These physiological characteristics were related to the needs of ecologists studying community energetics during changing seasonal temperatures.

Studies by Beames and Lindeborg (16) indicated that the freshwater snail, *Physa anatina*, underwent capacity and resistance adaptation in response to temperature increases. River snails (taken from 26°C) had significantly higher oxygen consumption rates at 35°C than pond snails (38° to 40°C). Conversely, pond snails had significantly lower oxygen consumption at 25°C than river snails. Lethal temperatures were slightly lower for river snails than for pond snails, 40° to 41°C as compared to 42° to 43°C.

Temperature acclimation of oxygen consumption rate in the freshwater snail, *Physa hawnii*, was identified by Daniels and Armitage (45). The mean oxygen consumption of snails acclimated to 5°, 10°, 20°, and 25°C increased at successively higher temperatures, as did the range and standard error of the means for all groups. Snails acclimated to 20° and 25°C exhibited relative constancy of consumption rate at intermediate temperatures. Changes in acclimation and activity reduced the effects of slow environmental temperature rise. Oxygen consumption and survival were low at 30°C, suggesting that this was near the lethal level for the species.

Mangum and Sassaman (119) established the temperature dependance of two discrete metabolic states, i.e., "active" and "resting," in the polychaete, *Diopatra cuprea*. Oxygen consumption Q_{10}'s were 1.77 for active metabolism and 1.73 for resting between 17.5° and 27.5°C. The activity states alternated spontaneously at a temperature-dependent rate. (There were 6 bursts/hr at 12.5°C and about 14 bursts/hr at 27.5°C.)

The enzymic mechanisms of metabolic temperature compensation were discussed by Somero (170). This important homeostatic phenomenon was due, in part, to temperature-dependent changes in enzyme-substrate (E-S) affinities, whereby increases in temperature through most of an organism's physiological temperature range were accompanied by immediate decreases in affinities. Evidence was presented that changes in E-S affinity also may be important in setting thermal tolerance limits for an organism. The relative roles of E-S affinity and activation energy changes in evolutionary adaptation were discussed.

Muscle and liver tissues from cold-acclimated rainbow trout (5°C) were shown by Dean (48) to oxidize acetate and palmitate more rapidly than warm-acclimated (18°C) trout at the same temperatures. Acetate was converted to CO_2 at rates proportional to environmental temperature, even at levels well above lethal high temperatures for the intact organism.

Malessa (117) investigated relative manifestations of thermal acclimation by red and white muscle in the eel, *Anguilla vulgaris*. Tissue temperature adaptation of terminal oxidative metabolism was more pronounced in red muscle. Total cytochrome oxidase activity of white muscle changed little with seasonal temperature acclimation. Aerobic metabolism and relative muscle weight increased in red muscle of cold-acclimated adults.

Thermal compensation of respiratory enzymes in tissues of goldfish was identified by Caldwell (36). Activities of cytochrome oxidase, succinate-cytochrome-C reductase, and NADH-cytochrome-C reductase were all higher in brain, gill, and muscle homogenates and in gill mitochondrial preparations from 10°C acclimated fish than from those acclimated to 30°C. The reverse was true for cytochrome C_1. Mitochondrial electron transport systems seemed to compensate to meet the changing ATP demands of new environmental temperatures.

The effect of temperature acclimation (8°, 15°, 25°, and 32°C) on acetylchlorine synthesis by goldfish brain homogenates was studied by Hebb *et al.* (73). The enzyme responsible for synthesis was stable at incubation temperatures higher than previous acclimation temperatures. Higher aclimation temperatures yielded higher enzyme activity. There were no differences in the kinetic properties of the enzyme between 29° and 45°C, but there were at lower incubation temperatures (15° to 25°C).

Gait (60) determined the temperature of thermal inactivation of a liver enzyme, β-galactosidase, from adult, female pink salmon (*Oncorhynchus gorbuscha*). At a low pH of 4, the presumed microenvironment of the enzyme, thermal inactivation was not apparent at 45.5°C, although it occurred at 56.5°C. At neutral pH, inactivation occurred above 20°C. Rat liver at neutral pH was not inactivated until warmed to 40°C.

Lagerspetz and Tirri (100) presented evidence for a "transmitter hypothesis" of temperature acclimation in poiklotherms on the basis of experiments with ciliary activity in gills and heart beat rate of the freshwater mussel, *Anodonta cygnea cellensis*. Results suggested that the primary mechanism of temperature acclimation was a temperature-induced change in the release and/or metabolism of transmitter substances concerned with nervous control of physiological functions.

Dean and Berlin (49) identified altered hepatocyte function in rainbow trout acclimated to warm (18°C) and cold (5°C) temperatures. There was a higher level of total liver protein (mg/g wet wt) and a higher level of incorporation of ^{14}C-leucine into total liver protein in the cold-acclimated trout. The secretory rate decreased in cold fish.

The effect of cycling temperatures (8° to 18°C) on electrolyte balance in skeletal muscle and plasma of rainbow trout (*Salmo gairdneri*) was studied by Toews and Hickman (184). In most cases plasma levels of sodium, potassium, and chloride did not exceed levels in control fish held at low temperatures (8° and 10°C), suggesting that cycled trout acclimated to the lowest temperatures of the cycle. There was a significant inverse relationship between muscle potassium and muscle sodium levels in the cycled fish that was not exhibited by control fish at constant temperatures within the cycle. Rao (152) determined that both osmotic and chloride concentrations of blood plasma from rainbow were higher at 5°C than at 15°C.

The cellular dynamics of intestinal epithelium in coho salmon were studied by Johnson (85) to determine effects of temperature and whole-body X-irradiation. The renewal of cells was dependent on temperature, with estimates of 13 to 15 days, 23 to 25 days, and more than 35 days for renewal in fish acclimated to 18°, 10°, and 5°C, respectively. A series of *in vitro* enzyme assays using radiotracers demonstrated that the activities of some precursors of DNA synthesis, thymidine kinase activities, and total protein per unit volume of intestinal microsomal supernatant fraction were dependent on acclimation temperature.

The body temperature of poikilotherms has come under recent scrutiny. Carey and Teal (37) determined that bluefin tuna, *Thunnus thynnus*, can control internal body temperature (25° to 30°C) so that the warmest muscles vary only 5°C over a 10° to 30°C range of water temperature. Temperature regulation in this "poikilotherm" was correlated with wide tolerance of natural thermal variations. Body temperatures of salmon (*Salmo salar*) undergoing strong exercise were found by Lyman (110) to approximate water temperature even though blood and water were not in equilibrium. The gills played the major role in rapid conduction of heat from body tissues but other surfaces were of little significance. No correlation between fish size and body temperature was evident.

Variations in blood elements and volumes of plasma and blood associated with the thermoacclimatory process in brook trout, *Salvelinus fontinalis*, were investigated by Houston and DeWilde (76). Warm-acclimated trout had larger numbers of variations, but fewer erythrocytes than cold-acclimated fish. Mean cellular hemoglobin was stable, but O_2 content increased due to changes in plasma and blood volume. The data were discussed in relation to metabolic compensation to high temperature. Smirnova (168) demonstrated that increases in the numbers of granular leucocytes circulating in fish blood were associated with temperature rise, electrical stimulation, exposure to organic toxicants, and other stresses. There were corresponding increases in disintegration of red blood cells.

Poston *et al.* (148) investigated the effects of water temperature on specific dietary fats and varying caloric ratios in brown trout, *Salmo trutta*. Supplemental calories increased growth at 12.4°C but not at 8.3°C. Conversion of food into flesh was more efficient at the warmer temperature. Supplemental oil in the feed induced more body fat at 12.4°C than at 8.3°C. There were also differences in growth and body chemistry associated only with diet.

The biological half-life ($T_{\frac{1}{2}}$) of the slow components of ^{137}Cs and ^{24}Na in fish were found by Häsänen *et al.* (70) to decrease by approximately half following a 10°C rise in temperature. For yearling rainbow trout, $T_{\frac{1}{2}}$ for ^{137}Cs increased from 20 days at 20°C to 36 days at 7° to 8°C. In 5-yr-old carp, it was 55 days at 20°C and 120 days at 10°C. At 20° and 10°C, respectively, $T_{\frac{1}{2}}$ values for ^{24}Na were 7 and 15 days for perch, 7 and 11 days for roach, and 10 and 25 days for carp.

Persistent temperature change was found by Peak *et al.* (143) to cause significant alterations in the size of various organs of the fish, *Tilapia sparrmanii*, and in the respiration rate of excised liver. No change occurred in water content of the liver. Prolonged thiourea treatment induced hyperplastic goiter and an increase in liver respiration, but did not inhibit adaptive responses of the liver to temperature nor compensatory changes in ventilation rate.

Concentrations of potassium and calcium in seawater were shown by Shea *et al.* (161) to affect responsiveness of peripheral nerves from the legs of *Callinectes sapidus* (blue crab) to cold. All thermal responsiveness was abolished above 30 mM/l of potassium, while increases from 15 mM/l progressively decreased the threshold for cold response.

Tett (183) investigated the effects of temperature on the flash-stimulated luminescence of the euphausid, *Thysanoessa raschi*. Response delay and response duration decreased with increasing temperature from 0° to 20°C.

The junctional resistance at septa of the lateral giant axon from crayfish was found by Payton *et al.* (142) to be related inversely to temperature, with a Q_{10} of about 3 from 5° to 20°C. Nonjunctional axonal membrane was less affected. Change in resistance occurred rapidly with temperature change, suggesting thermal effects on membrane structure.

Kogan *et al.* (96) demonstrated the ability of thermophilic Actinomycetes, belonging to the genera *Actinomyces, Thermopolyspora, Micromonospora,* and *Actinobifiola* to transform several biochemical compounds.

Disease

The effect of the bacterial pathogen *Gaffkya homari* on the American lobster, *Homarus americanus*, was found by Stewart *et al.* (175) to vary with temperature. The mean time to death following a standard innoculum was 2 days at 20°C, 12 days at 15°C, 28 days at 10°C, 65 days at 7°C, 84 days at 5°C, and 172 days at 3°C. No deaths from the pathogen occurred at 1°C, but fatal infections resulted when the environmental temperature of these animals was raised subsequently.

Four papers dealt with temperature and parasitism. Vernberg and Vernberg (192) demonstrated that the thermal acclimation pattern of cytochrome-C oxidase from the digestive gland of the snail, *Nassarius obsoleta*, as well as the rate of activity, was altered by infections of the larval trematode *Zoogonus lasius*. Sporocysts and infected and noninfected snail tissues first acclimated to 10° or 25°C and then tested at 10°, 15°, 20°, 25°, 30°, 35°, and 40°C revealed distinctive Q_{10} patterns. Enzymatic activity of the parasite was lower than the host tissues at each assay temperature, and the pattern of thermal acclimation was different.

Incidence of the monogenetic trematode *Gyrodactylus elegans* on the epidermis of goldfish was found by Anthony (8) to decrease as the water warmed either seasonally or by artificial heating. Seasonal abundance peaked in April and May when monthly mean water temperatures were 9° and 11°C. There was a relative increase in incidence on the gills at higher temperatures.

Ko and Adams (95) studied larval development of a parasitic nematode, *Philonema oncorhynchi*, in a copepod intermediate host (*Cyclops bicuspidatus*) at 4°, 10°, and 15°C. The size of the ultimate infective stage (for salmon) was unaffected by temperature. Development was directly proportional to temperature, in which a 5°C increase halved the rate. The ecological significance of the findings to natural infections in juvenile sockeye salmon were discussed.

Complete development of the filarial parasite, *Dirofilaria immitis*, in adult mosquitoes, *Aedes aegyptii*, was found by Singh *et al.* (164) to occur at 20°, 25°, and 30°C. The most favorable range was between 25° and 30°C. A temperature of 35°C was detrimental both to filarial development and to survival of the mosquitoes.

Two papers reported the effects of temperature on antibody formation in fish. Chiller (38) found that low acclimation temperatures (6° to 8°C) markedly suppressed the immune response of antigenized rainbow trout. Transfer of hyperantigenized, cold-acclimated fish to warmer water made the response manifest. Optimal plaque numbers developed on Jerne plates during 4-hr incubation periods at 18°C, a temperature close to the summertime temperature of the trout's natural environment. Muroga and Equsa (130) found the maximum rate of titer development against the bacterium *Vibrio anguillarum*, in the Japanese eel, *Anguilla japonica*, to occur between 15° and 23°C, while no difference in rate was found between 23° and 27°C. Similarly high titers eventually were attained at all temperatures between 15° and 27°C, but no measurable antibody was produced at a constant 11°C. Titers rose below 11°C if antibody production had started at a higher temperature.

Primary Production

Trukhin and Mikryakova (188) investigated growth of *Chlorella* cultures for several days at 25°, 32°, and 39°C and found that the dry weight produced was independent of culture temperature at nonsaturating light intensities for photosynthesis. Respiration rate of the algae did not increase with elevated temperature, but adapted rapidly to the change. Compensatory changes in respiratory enzyme activities with temperature change appeared to be the mechanism.

Keller *et al.* (88) investigated proliferation of *Chlorella* in cultures associated with simultaneous changes in three variables—temperature (15° to 33°C), salinity (0 or 0.6 percent), and bacteria (present or absent). Temperature dependence of algal growth was exhibited strongly by bacteria-free cultures with zero salinity, but poorly in other cultures corresponding to natural estuarine conditions.

The combined effects of temperature and light intensity on *in vitro* growth of marine algae were investigated by Yanase and Imai (200). Optimum conditions for *Monochrysis lutheri* were 4,500 to 8,000 lux at 23° to 25°C; for *Platymonas sp.*, 4,500 to 8,000 lux at 23° to 25°C; for *Nitzschia closterium*, 4,500 to 12,500 lux at 23° to 29°C; and for *Chaetoceros calcitrans*, 4,500 to 12,500 lux at 23°C. The fastest growing algae under optimum conditions was *C. calcitrans*.

Slobodskoi *et al.* (166) derived analytical expressions for determining temperature dependence of production by dense cultures of microalgae. Optimum temperatures for growth were presented, along with comparisons of experimental and calculated data.

The effects of light and temperature on photosynthetic activities of cultured unicellular green algae (*Dunaliella salina* and *Platymonas* sp.) and diatoms (*Nitzschia kuetzingiana* and *N. ovalis*) were studied by Fedorov *et al.* (59). Rate of CO_2 fixation per unit biomass under varying conditions of light and temperature differed considerably among the species. A notable increase in primary productivity of *D. salina* occurred only when both variables were raised simultaneously, while the diatoms reacted positively to temperature rise alone (between 10° and 20°C). Activity by *Platymonas* decreased with temperature rise at 200 lux, but increased at 4,000 lux.

Photosynthesis of thermal algae (mostly *Synechococcus spp.*) in a natural spring was studied by Brock and Brock (26). Self-shading was extensive in thick algal mats that reduced both photosynthesis and responses to light changes by the entire biomass. Photosynthesis fell progressively with decreasing light, although the most efficient use of available light occurred at 7 to 14 percent of full sunlight. Chlorophyll content of algae at the mat surface increased markedly in response to light attenuation.

The recovery of an algal mat destroyed by a violent hailstorm at Mushroom Spring (thermal) in Yellowstone National Park was studied by Brock and Brock (25). Doubling of cell numbers at three stations occurred in 17 days (71°C), 10.5 days (68°C), and 10 days (65°C). These rates were somewhat less than those reported previously for steady-state communities at the same locations. Normal mat size was attained 152 days after destruction.

Stockner (176) studied the ecology of a diatom community in a thermal stream at Mt. Rainier National Park that was characterized by stable temperatures (33.6° to 35°C) and uniform chemical composition. Diatom species fluctuated cyclically in abundance over the 3-yr study period. Species diversity and redundancy were quite uniform, suggesting community stability despite shifts in species composition. Core samples indicated gradual community evolution following drastic environmental changes about 500 yr ago.

Feeding by adult and larval flies (*Paraloenia turbida* and *Ephydra bruesi*) on algal and bacterial mats in hot springs was identified by Brock *et al.* (28) by means of radiocarbon techniques. Intense grazing by flies accounted for discrepancies noted in earlier studies between standing crop of algae and primary production in zones with temperatures of 30° to 45°C.

The influence of 116 combinations of temperature (2°, 7°, 12°, 16°C), salinity (5 to 35 ppt at 5 ppt intervals), and light (5 levels) on the mean daily cell division rate (K) of the marine diatom, *Detonula confervacea*, from Narragansett Bay was examined by Smayda (167). No growth occurred at 16°C, but growth proceeded under favorable light and salinity at 2°, 7°, and 12°C. An increased optimal light intensity occurred at increased temperature 200 to 600 ft-c (2,160 to 6,490 lumen/sq m) at 2°C, 600 to 1,200 ft-c (6,490 to 12,980 lumen/sq m) at 7°C, and 1,200 to 1,800 ft-c (12,980 to 19,470 lumen/sq m) at 12°C. At 32 ppt salinity and 1,100 to 1,200 ft-c, K increased 2.5-fold from 0.6 to 1.5 between 2° and 12°C. The optimal salinity range, 15 to 30 ppt, was independent of temperature.

Spore germination in the green alga, *Ulva pertusia*, was found by Masao and Arasaki (120) to occur over a wide range of temperature (3° to 32°C). The optimal range was about 20° to 25°C. Growth was more rapid at higher light intensities but was restricted by temperature at about 25°C. Survival of spores in darkness was improved at low temperatures (5° to 10°C).

The movement of cytoplasm and chloroplasts in cells of the aquatic plant *Elodea* at various temperatures was shown by Tageeva and Kazantsev (180) to be a "probability process" characterized by a "velocity spectrum." The relationship between movements of cytoplasm and chloroplasts varied with temperature. The activity of an actinomyosin-like plant protein was suggested.

Stepanskii and Yaglova (173) demonstrated the temperature dependence of membrane potentials of internodal cells of the alga, *Nitella mucronata*. The temperature range for reversible effects and the importance of the cooling rate for reversibility were determined. There was a distinct seasonal variability to the results. Thermal coagulation of cytoplasmic proteins and K^+ and Na^+ ion imbalances were proposed mechanisms.

Temperature shocks of 14.5° to 28.5°C did not damage heterotrophic cultures of the flagellate, *Euglena gracilis*, according to Neal *et al.* (134). The temperature change induced synchronous cell division during the warm phase which was maintained with alternating periods of warm and cold. Kirschstein (94) found a temperature-sensitive, circadian, rhythmic mobility in a colorless, heterotrophic mutant of *E. gracilis* held in darkness. Sensitivity apparently was compensated by diurnal photosynthesis in mixotrophic cells. The cycling of mobility was dependent on ambient temperature during anaerobic glycolysis and independent of temperature during respiration, just as in green forms.

Nitrogen fixation in thermal waters was reported in two papers. An aerobic, nitrogen-fixing bacterium, identified as *Pseudomonas ambigua*, was isolated from thermal waters in the Kurile Islands by Golovacheva and Kalininskaya (61). Nitrogen-fixation in this species was heretofore unrecognized. Its efficiency on an artificial medium was 4.2 to 4.6 mg N/g of utilized glucose. *Azotobacter chroococcum* and *A. agilis* exhibited growth and fixation of molecular nitrogen at 30° and 35°C when cultured by Kalininskaya and Golovacheva (86). Growth, but not N-fixation, was inhibited at 40°C. Nitrogen fixation by *Pseudomonas ambigua* was most effective at 30°C.

Waste Assimilation

The effects of temperature on biological oxidation processes in waste treatment were studied by Shih and Stack (162). Oxygen consumption per unit of COD varied with substrates oxidized and temperature. The predominant biological populations and their growth rates changed with temperature, as did production of relatively nondegradable materials. Temperature-acclimated biological growth was required for reliable measurements of oxygen consumption.

Krenkel *et al.* (99) reviewed the effects of temperature increases on waste assimilation capacities of streams and reservoirs. While the reaeration coefficient for rivers or impoundments rose slowly with increased temperature, the deoxygenation coefficient (i.e., rate of oxygen use by microorganisms and other biota) rose rapidly. Temperature increases were equivalent to adding additional oxygen-demanding organic material to the water (e.g., at 25°C, a 5°C increase = 11,000 lb BOD/day in the free-flowing Coosa River, Ga.). The same information also was presented by Krenkel and Parker (98). These results were discussed by Edinger (53), who presented other oxygen-sag modeling data that suggested over-assessment of reduction in waste assimilation capacity by Krenkel and Parker.

MacKay and Fleming (114) correlated DO levels in the polluted Clyde Estuary, Scotland, with freshwater inflows and water temperature. A simple equation was prepared that linked three parameters at 13 sampling stations. DO levels (as percent saturation) generally increased with greater freshwater inflow and decreased with rise in temperature.

Beneficial Uses

Three papers described attempts at fish culture in artificially warmed water. Nash (133) demonstrated the biological feasibility of rearing marine fish in the warm-water effluent of electrical generating stations. Plaice (*Pleuronectes platessa*) and sole (*Solea vulgaris*) were raised to marketable size at the Hunterstan, Scotland, nuclear power station in half the time required in the natural environment. Many problems remained to be solved before a commercial operation is feasible, however.

Gribanov *et al.* (64) described the culture of carp in tanks placed in the cooling pond of an electric power plant. Although there was no comparison with productivity in nonthermal water, a stocking rate of 250 fish/sq m produced 100 kg/sq m of 400 g carp at the end of the rearing period. Menzel (123) discussed similar industrial production of carp in warm-water ponds. Problems of optimizing fry production and supply, reducing feeding costs, and reducing manpower needs remain to be resolved. Theoretically, there exists a great potential for exceedingly high protein yields.

References

1. Abbott, F. S., "The Effects of Light and Temperature on Isolated Melanophores in *Fundulus heteroclitus* L." *Can. Jour. Zool.*, **47**, 203 (1969).
2. Adams, J. R., "Thermal Power, Aquatic Life, and Kilowatts on the Pacific Coast." *Nuclear News*, **75** (Sept. 1969).
3. Aiken, D. E., "Ovarian Maturation and Egg Laying in the Crayfish *Orconectes virilis:* Influence of Temperature and Photoperiod." *Can. Jour. Zool.*, **47**, 931 (1969).
4. Alabaster, J. S., "Effects of Heated Discharges on Freshwater Fish in Britain." In "Biological Aspects of Thermal Pollution." Krenkel, P. A., and Parker, F. L., Eds., Vanderbilt Univ. Press, Nashville, Tenn., 354 (1969).
5. Allen, J. F., "Research Needs for Thermal-Pollution Control." In "Biological Aspects of Thermal Pollution." Krenkel, P. A., and Parker, F. L., Eds., Vanderbilt Univ. Press, Nashville, Tenn., 382 (1969).
6. Altukhov, K. A., Ben'ko, K. I., and Bulatovich, M. A., "Acclimatization of Rainbow Trout and Peled in the Carp Ponds of the Western Ukraine." *Prob. in Ichthyol.* (USSR), Amer. Fish. Soc. Transl., **8**, 726 (1968, 1969).
7. Anderson, J. F., "Influence of Photoperiod and Temperature on the Induction of Diapause in *Aedes atropalpus* (Diptera: Culicidae)." *Entomol. Exp. Appl.*, 110. 321 (1968); *Biol. Abs.*, **50**, 56146 (1969).
8. Anthony, J. D., "Temperature Effect on the Distribution of *Gyrodactylus elegans* on Goldfish." *Bull. Wildlife Disease Assoc.*, **5**, 44 (1969).
9. Averett, R. C., "Influence of Temperature on Energy and Material Utilization by Juvenile Coho Salmon." Ph.D. Dissertation, Oregon State Univ., Corvallis, Ore. (1969); *Dissertation Abs.*, **29**, 4435-B (1969).
10. Bachmann, K., "Temperature Adaptations of Amphibian Embryos." *Amer. Nat.*, **103**, 115 (1969).
11. Bai, A. R. K., Srihari, K., Shadaksharaswamy, M., and Jyothy, P. S., "The Effects of Temperature on *Blepharisma intermedium*." *Jour. Protozool.*, **16**, 738 (1969).
12. Baig, I. A., and Hopton, J. W., "Psychrophilic Properties and the Temperature Characteristic of Growth of Bacteria." *Jour. Bacteriol.*, **100**, 552 (1969).
13. Banks, J. W., "A Review of the Literature on the Upstream Migration of Adult Salmonids." *Jour. Fish. Biol.* (Brit.), **1**, 85 (1969).
14. Barnes, H., and Barnes, M., "Seasonal Changes in the Acutely Determined Oxygen Consumption and Effect of Temperature for Three Common Cirripedes, *Balanus balanoides* (L.), *B. balanus* (L.), and *Chthamalus stellatus* (Poli)." *Jour. Exp. Mar. Biol. Ecol.*, **4**, 36 (1969).
15. Battelle-Northwest, "Biological Effects of Thermal Discharges: Annual Progress Report for 1968." Reprinted from Pacific Northwest Laboratory Annual Report for 1968 to U.S.A.E.C. Division of Biology and Medicine. Vol. I: Life Sciences. U.S.A.E.C. Res. and Dev. Rept. No. BNWL-1050. (Battelle-Northwest, Richland, Washington) 49 p.
16. Beames, C. G., Jr., and Lindeborg, R., "Temperature Adaptation in the Snail *Physa anatina*." *Proc. Okla. Acad. Sci.*, **48**, 12 (1969).
17. Beer, L. P., and Pipes, W. O., "A Practical Approach to the Preservation of the Aquatic Environment: The Effects of Discharge of Condenser Water into the Mississippi River." Commonwealth Edison Co., Chicago, Ill. (1969).
18. Berenbeim, D. Ya., "Calculating the Correlation Coefficient Between the Mean Annual Temperature of the Water at Spawning Grounds and the Spawning Times of Some Subspecies of Herring of the Pacific and Atlantic Oceans." *Tr. Atl. Nauch-Issled. Inst. Ryb. Khoz. Okeanogr.* (USSR), **18**, 215 (1967); *Biol. Abs.*, **50**, 45542 (1969).
19. Bidgood, B. F., and Berst, A. H., "Lethal Temperatures for Great Lakes Rainbow Trout." *Jour. Fish. Res. Bd. Can.*, **26**, 456 (1969).
20. Biernacka, I., "Influence of Salinity and of Temperature on Protists in the Baltic Sea and in the Vistula Lagoon." *Ekol. Pol.* (Poland), **16**, 262 (1968); *Biol. Abs.*, **50**, 107494 (1969).
21. Billard R., "Influence of Temperature on the Duration and Efficacy of the Spermatogenesis of the Guppy, *Poecilla reticulata*." *C. R. Hebd. Seances Acad. Sci., Ser. D. Sci. Natur* (France), **266**, 2287 (1968); *Biol. Abs.*, **50**, 125073 (1969).

22. Bott, T. L., and Brock, T. D., "Bacterial Growth Rates Above 90°C in Yellowstone Hot Springs." *Science*, **164**, 1411 (1969).

23. Brenko, M. H., and Calabrese, A., "The Combined Effects of Salinity and Temperature on Larvae of the Mussel *Mytilus edulis*." *Marine Biol.* (West Germany), **4**, 224 (1969).

24. Brett, J. R. Shelbourn, J. E., and Shoop, C. T., "Growth Rate and Body Composition of Fingerling Sockeye Salmon, *Oncorhynchus nerka*, in Relation to Temperature and Ration Size." *Jour. Fish. Res. Bd. Can.*, **26**, 2363 (1969).

25. Brock, T. D., and Brock, M. L., "Recovery of a Hot Spring Community from a Catastrophe." *Jour. Phycol.*, **5**, 75 (1969).

26. Brock, T. D., and Brock, M. L., "Effect of Light Intensity on Photosynthesis by Thermal Algae Adapted to Natural and Reduced Sunlight." *Limnol. Oceanogr.*, **14**, 334 (1969).

27. Brock, T. D., and Freeze, H., "*Thermus aquaticus* gen. n. and sp. n., a Nonsporulating Extreme Thermophile." *Jour. Bacteriol.*, **98**, 289 (1969).

28. Brock, M. L., Wiegert, R. G., and Brock, T. D., "Feeding by *Paracoenia* and *Ephydra* (Diptera: Ephydridae) on the Microorganisms of Hot Springs." *Ecology*, **50**, 192 (1969).

29. Brown, R. F., Wildman, J. D., and Eppley, R. M., "Temperature-Dose Relationships with Aflatoxin on the Brine Shrimp, *Artemia salina*." *Jour. Ass. Offic. Anal. Chem.*, **51**, 905 (1968); *Biol. Abs.*, **50**, 42162 (1969).

30. Buchner, H., and Kiechle, H., "Effect of Temperature on Bisexuality in *Asplanchna*." *Naturwissenschaften* (Germany), **53**, 708 (1966); *Biol. Abs.*, **50**, 27813 (1969).

31. Buffington, J. D., "Temperature Acclimation of Respiration in *Culex pipiens pipiens* (Diptera: Culicidae) and the Influence of Seasonal Selection." *Comp. Biochem. Physiol.*, **30**, 865 (1969).

32. Burns, C. W., "Relation Between Filtering Rate, Temperature, and Body Size in Four Species of *Daphnia*." *Limnol. Oceanogr.*, **14**, 693 (1969).

33. Cairns, J., Jr., "Rate of Species Diversity Restoration Following Stress in Freshwater Protozoan Communities." *Sci. Bull.*, Univ. of Kansas, Lawrence, Kans., **48**, 209 (1969).

34. Cairns, J., Jr., "The Effects of Heated Discharges on Freshwater Benthos: Discussion." In "Biological Aspects of Thermal Pollution." Krenkel, P. A., and Parker, F. L., Eds., Vanderbilt Univ. Press, Nashville, Tenn., 214 (1969).

35. Calderon, E. G., "The Cultivation of Common and Rainbow Trout in High Temperature Waters." *An. Inst. Forest Invest. Exp. Madrid*, **1967**, 145 (1967); *Biol. Abs.*, **50**, 124183 (1969).

36. Caldwell, R. S., "Thermal Compensation of Respiratory Enzymes in Tissues of the Goldfish (*Carassius auratus* L.)." *Comp. Biochem. Physiol.*, **31**, 79 (1969).

37. Carey, F. G., and Teal, J. M., "Regulation of Body Temperature by the Blue Fin Tuna." *Comp. Biochem. Physiol.*, **28**, 205 (1969).

38. Chiller, J. M., "Studies of Antibody Formation in Rainbow Trout (*Salmo gairdneri*)." Ph.D. Dissertation, Univ. of Washington, Seattle, Wash. (1968); *Dissertation Abs.*, **29**, 2546-B (1969).

39. Churchill, M. A., and Wojtalik, T. A., "Effects of Heated Discharges: the TVA Experience." *Nuclear News*, **80** (Sept. 1969).

40. Clark, J. R., "Thermal Pollution and Aquatic Life." *Sci. American*, **220**, 18 (1969).

41. Coe, M. J., "Local Migration of *Tilapia grahami* Boulenger in Lake Magadi, Kenya in Response to Diurnal Temperature Changes in Shallow Water." *East. African Wildlife Jour.*, **5**, 171 (1967).

42. Cook, H. L., and Murphy, M. A., "The Culture of Larval Penaeid Shrimp." *Trans. Amer. Fish. Soc.*, **98**, 751 (1969).

43. Coutant, C. C., "Thermal Pollution—Biological Effects. A Review of the Literature of 1968." BNWL-2376, Battelle-Northwest, Richland, Wash.; *Jour. Water Poll. Control Fed.*, **41**, 1036 (1969).

44. Crippa, T. F., "Further Studies on Acquired Resistance to Temperature in *Paramecium aurelia* Syngen 1." *Boll. Inst. Biol. Univ. Genova Sez. Biol. Anim.* (Italy), **35**, 19 (1967); *Biol. Abs.*, **50**, 129005 (1969).

45. Daniels, J. M., and Armitage, K. B., "Temperature Acclimation and Oxygen Consumption in *Physa hawnii* Lea (Gastropoda: Pulmonata)." *Hydrobiologia* (Denmark), **33**, 1 (1969).

46. Davis, H. C., and Calabrese, A., "Survival and Growth of Larvae of the European Oyster (*Ostrea edulis* L.) at Different Temperatures." *Biol. Bull.*, **136**, 193 (1969).

47. Davis, W. S., "Conditions for Coexistence of Aquatic Communities with the

Expanding Nuclear Power Industry.'' *Nuclear Safety*, **10**, 292 (1969).

48. Dean, J. M., ''The Metabolism of Tissues of Thermally Acclimated Trout (*Salmo gairdneri*).'' *Comp. Biochem. Physiol.*, **29**, 185 (1969).

49. Dean, J. M., and Berlin, J. D., ''Alterations in Hepatocyte Function of Thermally Acclimated Rainbow Trout (*Salmo gairdneri*).'' *Comp. Biochem. Physiol.*, **29**, 307 (1969).

50. Doudoroff, P., ''Developing Thermal Requirements for Freshwater Fishes: Discussion.'' In ''Biological Aspects of Thermal Pollution.'' Krenkel, P. A., and Parker, F. L., Eds., Vanderbilt Univ. Press, Nashville, Tenn. 140 (1969).

51. Dow, R. L., ''Cyclic and Geographic Trends in Seawater Temperature and Abundance of American Lobster.'' *Science*, **164**, 1060 (1969).

52. Dunnington, E. A., ''Survival Time of Oysters After Burial at Various Temperatures.'' *Proc. Nat. Shellfish Assoc.*, **58**, 101 (1968).

53. Edinger, J. E., ''Impoundment and Temperature Effect on Waste Assimilation: Discussion.'' *Jour. San. Eng. Div., Proc. Amer. Soc. Civil Engr.*, **95**, 991 (1969).

54. Edwards, R. R. C., Finlayson, D. M., and Steele, J. H., ''The Ecology of O-Group Plaice and Common Dabs in Loch Ewe. II. Experimental Studies of Metabolism.'' *Jour. Exp. Mar. Biol Ecol.* (Neth.), **3**, 1 (1969).

55. Egami, N., ''Kinetics of Recovery from Injury After Whole-Body X-Irradiation of the Fish *Oryzias latipes* at Different Temperatures.'' *Radiat. Res.*, **37**, 192 (1969).

56. Ellis, R. A., and Borden, I. H., ''Effects of Temperature and Other Environmental Factors on *Notonecta undulata* Say (Hemiptera: Notonectidae).'' *Pan. Pac. Entomol.*, **45**, 20 (1969); *Biol. Abs.*, **50**, 84115 (1969).

57. European Inland Fisheries Advisory Commission Working Party on Water Quality Criteria for European Freshwater Fish, ''Water Quality Criteria for European Freshwater Fish—Water Temperature and Inland Fisheries.'' *Water Res.* (Brit.), **3**, 645 (1969).

58. Fagetti, E. G., ''Larval Development of the Spider Crab *Pisoides edwardsi* (Decapoda, Brachyura) Under Laboratory Conditions.'' *Marine Biol.* (West Germany), **4**, 160 (1969).

59. Fedorov, V. D., Maksimov, V. N., and Khromov, V. M., ''Effect of Light and Temperature on the Primary Production of Some Unicellular Green and Diatom Algae.'' *Fiziol. Rast.* (USSR), **15**, 640 (1968); *Biol. Abs.*, **50**, 83328 (1969).

60. Gatt, S., ''Thermal Lability of Beta Galactosidase from Pink Salmon Liver.'' *Science*, **164**, 1422 (1969).

61. Golovacheva, R. S., and Kalininskaya, T. A., ''*Pseudomonas ambigua*—A Nitrogen-Fixing Bacterium from Thermal Springs.'' *Mikrobiologia* (USSR), **37**, 941 (1968).

62. Gonor, J. J., ''Temperature Relations of Central Oregon Marine Intertidal Invertebrates: A Prepublication Technical Report to the Office of Naval Research.'' Dept. of Oceanography, Oregon State Univ., Corvallis, Ore. (1968).

63. Gorodilov, Yu. N., ''Study of the Sensitivity of Fish to High Temperature During Embryogenesis. I. Changes in Sensitivity to High Temperature Effects of Developing Eggs of Autumn Spawning Fish Species.'' *Tsitologiya* (USSR), **11**, 169 (1969); *Biol. Abs.*, **50**, 87765 (1969).

64. Gribanov, L. V., Korneev, A. N., Korneeva, L. A., and Pronin, G. M., ''Some Aspects of Carp Breeding and Nutrition in Tanks in Thermal Waters.'' *Tr. Vses. Nauch. Issled. Inst. Prudovogo Ryb. Khoz.* (USSR), **15**, 3 (1967); *Biol. Abs.*, **50**, 11842 (1969).

65. Grigg, G. C., ''Temperature-Induced Changes in the Oxygen Equilibrium Curve of the Blood of the Brown Bullhead, *Ictalurus nebulosus*.'' *Comp. Biochem. Physiol.*, **28**, 1203 (1969).

66. Haefner, P. A., Jr., ''Temperature and Salinity Tolerance of the Sand Shrimp, *Crangon septemspinosa* Say.'' *Physiol. Zool.*, **42**, 388 (1969).

67. Hammel, H. T., Strømme, S. B., and Myhre, K., ''Forebrain Temperature Activates Behavioral Thermoregulatory Response in Arctic Sculpins.'' *Science*, **165**, 83 (1969).

68. Harrington, R. W., ''Delimitation of the Thermolabile Phenocritical Period of Sex Determination and Differentiation in the Ontogeny of the Normally Hermaphroditic Fish *Rivulus marmoratus* Poey.'' *Physiol. Zool.*, **41**, 447 (1968).

69. Harvey, R. S., ''Effects of Temperature on the Sorption of Radionuclides by a Blue-Green Alga.'' *Proc. Second Nat. Symp. Radioecology*, Ann Arbor, Mich., May, 1967, 266 (1969).

70. Häsänen, E., Kolehmainen, S., and Miettinen, J. K., ''Biological Half-Times of Caesium-137 and Sodium-22 in Different Fish Species and Their Temperature Dependence.'' *Proc. 1st Int.*

Congr. Radiat. Prot., Rome, 1966, 401 (1968); *Water Poll. Abs.* (Brit.), **42**, 1720 (1969).

71. Haschemeyer, H. E. V., "Oxygen Consumption of Temperature-Acclimated Toadfish, *Opsanus tau*." *Biol. Bull.*, **136**, 28 (1969).

72. Hawkes, H. A., "Ecological Changes of Applied Significance Induced by the Discharge of Heated Waters." In: "Engineering Aspects of Thermal Pollution." Parker, F. L., and Krenkel, P. A., Eds., Vanderbilt Univ. Press, Nashville, Tenn., 15 (1969).

73. Hebb, C., Morris, D., and Smith, M. W., "Choline Acetyltransferase Activity in the Brain of Goldfish Acclimated to Different Temperatures." *Comp. Biochem. Physiol.*, **28**, 29 (1969).

74. Hedgpeth, J. W., and Gonor, J. J., "Aspects of the Potential Effect of Thermal Alteration on Marine and Estuarine Benthos." In "Biological Aspects of Thermal Pollution." Krenkel, D. A., and Parker, F. L., Eds., Vanderbilt Univ. Press, Nashville, Tenn., 80 (1969).

75. Heubach, W., "*Neomysis awatschensis* in the Sacramento-San Joaquin River Estuary." *Limnol. Oceanogr.*, **14**, 533 (1969).

76. Houston, A. H., and DeWilde, M. A., "Environmental Temperature and the Body Fluid System of the Freshwater Teleost: III. Hematology and Blood Volume of Thermally Acclimated Brook Trout, *Salvelinus fontinalis*." *Comp. Biochem. Physiol.*, **28**, 877 (1969).

77. Hurley, D. A., and Woodall, W. L., "Responses of Young Pink Salmon to Vertical Temperature and Salinity Gradients." *Int. Pac. Salmon Fish. Comm.*, Prog. Rept. No. 19 (1968).

78. Hussein, M. F., Badir, N., and Boulos, R., "Studies on the Effect of Insecticides as an Ecologically Induced Limiting Factor to the Life of Some Fresh Water Fish: III. Effect of Temperature on Toxaphene Toxicity to *Gambusia* sp. and *Tilapia zilli*." *Zool. Soc. Egypt Bull.*, **21**, 22 (1966/1967); *Biol. Abs.*, **50**, 103784 (1969).

79. Hyder, M., "Gonadal Development and Reproductive Activity of the Cichlid Fish *Tilapia leucosticta* (Trewavas) in an Equatorial Lake." *Nature* (Brit.), **224**, 1112 (1969).

80. Ingalls, T. H., Philbrook, F. R., and Majima, A., "Conjoined Twins in Zebra Fish." *Arch. Environ. Health*, **19**, 344 (1969).

81. Jankowsky, H. D., "Experiments on the Adaptation of Fish Within The Normal Range of Temperature." *Helgolander Wiss. Meeresunters.* (Germany), **18**, 317 (1968); *Biol. Abs.*, **50**, 42260 (1969).

82. Jaske, R. T., "The Need for Advance Planning of Thermal Discharges." *Nuclear News*, **65** (Sept. 1969).

83. Job, S. V., "The Respiratory Metabolism of *Tilapia mossambia* (Teleostei). II. The Effect of Size, Temperature, Salinity and Partial Pressure of Oxygen." *Marine Biol.* (West Germany), **3**, 222 (1969).

84. Johansen, P. H., "The Role of the Pituitary of the Goldfish in the Development of Resistance to Repeated High Temperature Exposures." *Can. Jour. Zool.*, **46**, 805 (1969).

85. Johnson, T. S., "Cellular Dynamics of the Intestinal Epithelium of Coho Salmon, *Oncorhynchus kisutch*, as Influenced by Temperature and X-rays." Ph.D. Dissertation, Oregon State Univ., Corvallis, Ore. (1969); *Dissertation Abs.*, **29**, 4442B (1969).

86. Kalininskaya, T. A., and Golovacheva, R. S., "Effect of Increased Temperatures on the Fixation of Molecular Nitrogen by Microorganisms from Thermal Zones." *Mikrobiologiya* (USSR), **38**, 316 (1969); *Biol. Abs.*, **50**, 93594 (1969).

87. Kalyuzhnaya, L. D., Kozhukhar, I. G., Bryanskaya, A. M., and Rogozhina, A. P., "Effect of Temperature on the Viability of Actinomycetes." *Microbiology* (USSR), **37**, 843 (1968).

88. Keller, E. C., Jr., Nagle, C. S., Jr., Keller, H. E., and Maxwell, D. C., "The Effect of Saline-Thermal-Bacterial Interactions on Populations of Primary Producers." *Proc. Penn. Acad. Sci.*, **41**, 97 (1968).

89. Kelly, M. T., and Brock, T. D., "Warm-Water Strain of *Leucothrix mucor*." *Jour. Bacteriol.*, **98**, 1402 (1969).

90. Kennedy, W. J., Taylor, J. D., and Hall, A., "Environmental and Biological Controls on Bivalve Shell Mineralogy." *Biol. Rev.*, **44**, 499 (1969).

91. Kenny, R., "Effects of Temperature, Salinity and Substrate on Distribution of *Clymenella torquata* (Leidy), Polychaeta." *Ecology*, **50**, 624 (1969).

92. Kenny, R., "Temperature Tolerance of the Polychaete Worms *Diopatra cuprea* and *Clymenella torquata*." *Marine Biol.* (West Germany), **4**, 219 (1969).

93. Khlebovich, V. V., and Lukanin, V. V., "Survival of the Spermatozoa of Certain White Sea Invertebrates in Water of Varying Salinity and Temperature." *Dokl. Akad. Nauk. SSSR* (USSR), **176**,

460 (1967); *Biol. Abs.*, **50**, 22285 (1969).

94. Kirschstein, M., "Rhythmic Behavior of a Colorless Mutant of *Euglena gracilis*." *Planta* (West Germany), **85**, 126 (1969); *Biol. Abs.*, **50**, 96162 (1969).

95. Ko, R. C., and Adams, J. R., "The Development of *Philonema oncorhynchi* (Nematoda: Philometridae) in *Cyclops bicuspidatus* in Relation to Temperature." *Can. Jour. Zool.*, **47**, 307 (1969).

96. Kogan, L. M., *et al.*, "Transformation of Steroids by Some Thermophilic Microorganisms." *Mikrobiologiya* (USSR), **37**, 628 (1968); *Biol. Abs.*, **50**, 54942 (1969).

97. Krenkel, P. A., and Parker, F. L. (Ed.), "Biological Aspects of Thermal Pollution." Vanderbilt Univ. Press, Nashville, Tenn. (1969).

98. Krenkel, P. A., and Parker, F. L., "Engineering Aspects, Sources, and Magnitude of Thermal Pollution." In "Biological Aspects of Thermal Pollution." Krenkel, P. A., and Parker, F. L., Eds., Vanderbilt Univ. Press, Nashville, Tenn., 10 (1969).

99. Krenkel, P. A., Thackston, E. L., and Parker, F. L., "Impoundment and Temperature Effect on Waste Assimilation." *Jour. San. Eng. Div., Proc. Amer. Soc. Civil Engr.*, **95**, SA1, 37 (1969).

100. Lagerspetz, K. Y. H., and Tirri, R., "Transmitter Substances and Temperature Acclimation in *Anodonta* (Pelecypoda)." *Ann. Zool. Fenn.* (Finland), **5**, 396 (1968); *Biol. Abs.*, **50**, 101120 (1969).

101. Lake, P. S., "The Effect of Temperature on Growth, Longevity and Egg Production in *Chirocephalus diaphanus* Prévost (Crustacea: Anostraca)." *Hydrobiologia* (Denmark), **33**, 342 (1969).

102. Lavatelli, G., Zuccarino, F., and Crippa, T. F., "Analysis of the Capacity of Resistance to Temperature in a Series of Strains of *Paramecium aurelia* Syngen 1." *Atti. Accad. Ligure Sci. Lett.* (Italy), **24**, 63 (1968); *Biol. Abs.*, **50**, 129105 (1969).

103. Leggett, W. C., and Power, G., "Differences Between Two Populations of Landlocked Atlantic Salmon (*Salmo salar*) in Newfoundland." *Jour. Fish. Res. Bd. Can.*, **26**, 1585 (1969).

104. Leicht, R., "The Oxygen Consumption of *Anguilla vulgaris* in its Relation to Temperature Differences Between Head Region and Body." *Marine Biol.* (West Germany), **3**, 28 (1969).

105. Levina, R. I., "Survival of Enterococcus (*Streptococcus faecalis*), *Salmonella typhosa* and *Escherichia coli* in River Water." *Gig. Sanit.* (USSR), **35**, 103 (1968); *Biol. Abs.*, **50**, 60409 (1969).

106. Lewis, K. B., and Jenkins, C., "Geographical Variation of *Nonionella flemingi*." *Micropaleontology*, **15**, 1 (1969); *Biol. Abs.*, **50**, 61515 (1969).

107. Lillelund, K., "Experiments on the Effects of Light and Temperature on Rearing Eggs of the Pike, *Esox lucius*." *Arch. Fisch. Wiss.* (Germany), **17**, 95 (1967); *Biol. Abs.*, **50**, 87776 (1969).

108. Loosanoff, V. L., "Maturation of Gonads of Oysters, *Crassostrea virginica*, of Different Geographical Areas Subjected to Relatively Low Temperatures." *Veliger*, **11**, 153 (1969).

109. Lowe, C. H., and Heath, W. G., "Behavioral and Physiological Responses to Temperature in the Desert Pupfish *Cyprinodon macularius*." *Physiol. Zool.*, **42**, 53 (1969).

110. Lyman, C. P., "Body Temperature of Exhausted Salmon." *Copeia*, **1968**, 631 (1968).

111. Lyutova, M. I., Feldman, N. L., and Drobyshev, V. P., "Changes in Cellular Thermoresistance of Marine Algae Due to Environmental Temperature." *Tsitologiya* (USSR), **10**, 1538 (1968); *Biol. Abs.*, **50**, 62321 (1969).

112. MacCrimmon, H. R., and Campbell, J. S., "World Distribution of Brook Trout, *Salvelinus fontinalis*." *Jour. Fish. Res. Bd. Can.*, **26**, 1699 (1969).

113. Macek, K. J., Hutchinson, C., and Cope, O. B., "The Effects of Temperature on the Susceptibility of Bluegills and Rainbow Trout to Selected Pesticides." *Bull. Env. Contam. Toxicol.*, **4**, 174 (1969).

114. MacKay, D. W., and Fleming, G., "Correlation of Dissolved Oxygen Levels, Freshwater Flows and Temperatures in a Polluted Estuary." *Water Res.* (Brit.), **3**, 121 (1969).

115. Maheshwari, R., "Occurrence and Isolation of Thermophilic Fungi." *Curr. Sci.* (India), **37**, 277 (1968); *Biol. Abs.*, **50**, 54992 (1969).

116. Makhlin, E. E., and Skholl, E. D., "Effect of Environmental Temperature on the Thermostability of Muscles, Glycerinated Muscle Tissue and Aldolase of Molluscs from the Barents Sea." *Tsitologiya* (USSR), **10**, 1442 (1968); *Biol. Abs.*, **50**, 134615 (1969).

117. Malessa, P., "Contributions to the Temperature Adaptation of the Eel (*Anguilla vulgaris*). III. Intensity and Distribution of the Activity of Succinate Dehydrogenase and Cytochrome Oxidase in the Lateral Muscle of Juvenile and Adult Animals." *Marine Biol.* (West Germany), 3, 143 (1969).

118. Mangum, C. P., "Low Temperature Blockage of the Feeding Response in Boreal and Temperate Zone Polychaetes." *Chesapeake Sci.*, 10, 64 (1969).

119. Mangum, C. P., and Sassaman, C., "Temperature Sensitivity of Active and Resting Metabolism in a Polychaetous Annelid." *Comp. Biochem. Physiol.*, 30, 111 (1969).

120. Masao, O., and Arasaki, S., "Physiological Studies on the Development of the Green Alga *Ulva pertusa*. I. Effect of Temperature and Light on the Development of Early Stage." *Rec. Oceanogr. Works Jap.*, 9, 129 (1967); *Biol. Abs.*, 50, 83397 (1969).

121. McRitchie, R. G., "Effects of Thermal and Osmotic Acclimation on an Estuarine Gastropod." Ph.D. Dissertation, Rice Univ., Houston, Tex. (1968); *Dissertation Abs.*, 29, 2297-B (1969).

122. Meldrim, J. W., "The Ecological Zoogeography of the Olympic Mudminnow (*Novumbra hubbsi* Schultz)." Ph.D. Dissertation, Univ. of Washington, Seattle, Wash. (1968); *Dissertation Abs.*, 30, 994-B (1969).

123. Menzel, H. D., "Estimate of Possible Accomplishments of Industrial Carp Production in Warm Water Discharges for a Managed Inland Fishery in the German Democratic Republic." *Z. Fischerei N. F.*, 17, 133 (1969).

124. Molander, A. R., and Swedmark, M. M., "Experimental Investigations on Variation in Plaice (*Pleuronectes platessa* Linne)." *Inst. Mar. Res. Lysekil.* (Sweden), 7, 3 (1967); *Biol. Abs.*, 50, 78691 (1969).

125. Moore, D. R., and Reish, D. J., "Studies on the *Mytilus edulis* Community of Alamitos Bay, California: IV. Seasonal Variation in the Gametes from Different Regions of the Bay." *Veliger*, 11, 250 (1969).

126. Morgan, P. V., and Bramer, H. C., "Thermal Pollution as Factor in Power Plant Site Selection." *Nuclear News*, 70 (Sept. 1969).

127. Mount, D. I., "Developing Thermal Requirements for Freshwater Fishes." In "Biological Aspects of Thermal Pollution." Krenkel, P. A., and Parker, F. L., Eds., Vanderbilt Univ. Press, Nashville, Tenn., 140 (1969).

128. Moyer, S., and Raney, E. C., "Thermal Discharges from Large Nuclear Plant." *Jour. San. Eng. Div., Proc. Amer. Soc. Civil Engr.*, 95, SA6, 1131 (1969).

129. Mukhin, A. I., "Effect of Gadoid Abundance and Water Temperature on the Trawl Fishery of the Southern Barents Sea." *Tr. Polyar Nauch Issled Proekt Inst. Morsk Ryb Khoz Okeanogr.*, (USSR), 20, 179 (1967); *Biol. Abs.*, 50, 11882 (1969).

130. Muroga, K., and Egusa, S., "Immune Response of the Japanese Eel to *Vibrio anguillarum*—I. Effects of Temperature on Agglutinating Antibody Production in Starved Eels." *Bull. Jap. Soc. Sci. Fish.*, 35, 868 (1969).

131. Nagabhushanam, R., and Sarojini, R., "Effect of Temperature and Salinity on the Heat Tolerance in the Hermit Crab, *Diogenes bicristimanus*." *Hydrobiologia* (Denmark), 34, 126 (1969).

132. Nakatani, R. E., "Effects of Heated Discharges on Anadromous Fishes." In "Biological Aspects of Thermal Pollution." Krenkel, P. A., and Parker, F. L., Eds., Vanderbilt Univ. Press, Nashville, Tenn., 294 (1969).

133. Nash, C. E., "Power Stations as Sea Farms." *New Scientist* (Brit.), 14, 367 (1968).

134. Neal, W. K., II, Funkhouser, E. A., and Price, C. A., "Large-Scale Temperature-Induced Synchrony of Cell Division in *Euglena gracilis*." *Jour. Protozool.*, 15, 761 (1969).

135. North, W. J., "Aspects of the Potential Effect of Thermal Alteration on Marine and Estuarine Benthos: Discussion." In "Biological Aspects of Thermal Pollution." Krenkel, P. A., and Parker, F. L., Eds., Vanderbilt Univ. Press, Nashville, Tenn., 119 (1969).

136. Ouchi, K., "Effects of Water Temperature on the Scale Growth and Width of the Ridge Distance in Goldfish." *Bull. Japan. Soc. Sci. Fish.*, 35, 25 (1969).

137. Pamatmat, M. M., "Seasonal Respiration of *Transennella tantilla* Gould." *Amer. Zoologist*, 9, 418 (1969).

138. Parker, F. L., and Krenkel, P. A., (Ed.), "Engineering Aspects of Thermal Pollution." Vanderbilt Univ. Press, Nashville, Tenn. (1969).

139. Parvatheswararao, V., "Initial Adaptive Responses to Thermal Stress in Freshwater Teleosts: I. Oxygen Consumption of Whole Animal." *Proc.*

Indian Acad. Sci. Sect. B., **68**, 225 (1968); *Biol. Abs.*, **50**, 98399 (1969).

140. Patrick, R., "Some Effects of Temperature on Freshwater Algae." In "Biological Aspects of Thermal Pollution." Krenkel, P. A., and Parker, F. L., Eds., Vanderbilt Univ. Press, Nashville, Tenn., 161 (1969).
141. Pattee, E., "Thermal Coefficients and Ecology of Some Fresh-Water Planarians. II. Tolerance of *Dugesia gonocephala.*" *Ann. Limnol.*, **4**, 99 (1968); *Biol. Abs.*, **50**, 89677 (1969).
142. Payton, B. W., Bennett, M. V. L., and Pappas, G. D., "Temperature Dependence of Resistance at an Electrotonic Synapse." *Science*, **165**, 594 (1969).
143. Peak, M. J., Fuller, N. R., and Peak, J. G., "Temperature Acclimation in *Tilapia sparrmanii:* Some Effects of Thiourea." *Zool. Afr.* (S. Africa), **3**, 95 (1967); *Biol. Abs.*, **50**, 30993 (1969).
144. Peterson, R. H., and Anderson, J. M., "Influence of Temperature Change on Spontaneous Locomotor Activity and Oxygen Consumption of Atlantic Salmon, *Salmo salar*, Acclimated to Two Temperatures." *Jour. Fish. Res. Bd. Can.*, **26**, 93 (1969).
145. Peterson, R. H., and Anderson, J. M., "Effects of Temperature on Brain Tissue Oxygen Consumption in Salmonid Fishes." *Can. Jour. Zool.*, **47**, 1345 (1969).
146. Petrovska, L., "The Microflora of the Thermal Springs of Vranjska Banja." *Fragm. Balcan. Mus. Macedonici Sci. Natur.*, **6**, 57 (1967); *Biol. Abs.*, **50**, 4008 (1969).
147. Poltoracka, J., "Specific Composition of Phytoplankton in a Lake Warmed by Waste Water from a Thermoelectric Plant and Lakes with a Normal Temperature." *Acta Soc. Bot. Pol.* (Poland), **37**, 297 (1968); *Biol. Abs.*, **50**, 23148 (1969).
148. Poston, H. A., Livingston, D. L., and Phillips, A. M., Jr., "The Effect of Source of Dietary Fat, Calorie Ratio, and Water Temperature on Growth and Chemical Composition of Brown Trout." *Fisheries Res.*, Rept. No. 32, Cortland (N. Y.) Hatchery, 14 (1969).
149. Precht, I., "Investigations on Diapause, Capacity-Adaptation and Temperature Resistance of Some Insects and Snails." *Zeits. Wiss. Zool.*, **176**, 122 (1967); Biol. Abs., **50**, 22287 (1969).
150. Proffitt, M. A., "Effects of Heated Discharge upon Aquatic Resources of White River at Petersburg, Indiana." Indiana Univ. Water Resources Res. Ctr., Repts. of Investigations No. 3 (1969).
151. Raney, E. C., and Menzel, B. W., "Heated Effluents and Effects on Aquatic Life with Emphasis on Fishes —A Bibliography." Cornell Univ. Wat. Resources and Marine Sci. Ctr., Philadelphia Electric Co., Philadelphia, Pa., and Ichthyol. Assoc., Bull. No. 2 (1969).
152. Rao, G. M. M., "Effect of Activity, Salinity and Temperature on Plasma Concentrations of Rainbow Trout." *Can. Jour. Zool.*, **47**, 131 (1969).
153. Reed, P. H., "Culture Methods and Effects of Temperature and Salinity on Survival and Growth of Dungeness Crab (*Cancer magister*) Larvae in the Laboratory." *Jour. Fish. Res. Bd. Can.*, **26**, 389 (1969).
154. Regan, L., "*Euphausia pacifica* and Other Euphausids in the Coastal Waters of British Columbia: Relationships to Temperature, Salinity and Other Properties in the Field and Laboratory." Ph.D. Dissertation, Univ. of British Columbia, Vancouver, B. C. (1968); *Dissertation Abs.*, **30**, 905-B (1969).
155. Reznichenko, P. N., Gulidov, M. V., and Kotlyarevskaya, N. V., "Survival of Eggs of the Tench *Tinca tinca* (L.) Incubated at Constant Temperatures." *Prob. in Ichthyol* (USSR), Amer. Fish. Soc. Translation, **8**, 391 (1968, 1969).
156. Round, F. E., "Light and Temperature: Some Aspects of Their Influence on Algae." In: "Algae, Man, and the Environment." Jackson, D. F., Ed., Syracuse Univ. Press, Syracuse, N. Y. (1968).
157. Savitz, J., "Effects of Temperature and Body Weight on Endogenous Nitrogen Excretion in the Bluegill Sunfish (*Lepomis macrochirus*)." *Jour. Fish. Res. Bd. Can.*, **26**, 1813 (1969).
158. Schaller, F., "Effect of Temperature on Diapause and Growth of the Embryo of *Aeschna mixta* (Odonata)." *Jour. Insect. Physiol.*, **14**, 1447 (1968).
159. Schneider, L., "Experimental Investigations on the Influence of Day Length and Temperature on Gonad Development in the Three-Spined Stickleback, *Gasterosteus aculeatus.*" *Oecologia* (Germany), **3**, 249 (1969).
160. Shabalina, A. A., "Effects of Cobalt Chloride on Physiological Indices in the Rainbow Trout (*Salmo iridens* Gibbons)." *Prob. in Ichthyol.* (USSR), Amer. Fish. Soc. Translation, **8**, 741 (1968, 1969).

161. Shea, S., Sigafoos, D., and Scott, D., Jr., "The Effect of Calcium and Potassium on the Thermal Excitability of a Model Thermoreceptor." *Comp. Biochem. Physiol.*, 28, 701 (1969).
162. Shih, C. S., and Stack, V. T., Jr., "Temperature Effects on Energy Oxygen Requirements in Biological Oxidation." *Jour. Water Poll. Control Fed.*, 41, R461 (1969).
163. Shrable, J. B., Tiemeier, O. W., and Deyoe, C. W., "Effects of Temperature on Rate of Digestion by Channel Catfish." *Prog. Fish-Cult.*, 31, 131 (1969).
164. Singh, D., Mammen, M. L., and Das, M., "Effect of Temperature on the Extrinsic Incubation of *Dirofilaria immitis* in *Aedes aegypti*." *Bull. Indian Soc. Malaria Commun. Dis.*, 4, 139 (1967); *Biol. Abs.*, 50, 112145 (1969).
165. Skoeb, Y., and Eger, M., "Restorative Effect of a Low Temperature on Survival, the Amount of RNA and Proteins of the Irradiated Amoeba." *C. R. Hebd. Seances Acad. Sci., Ser. D. Sci. Natur.* (France), 264, 477 (1967); *Biol. Abs.*, 50, 89822 (1969).
166. Slobodskoi, L. I., *et al.*, "Analytical Expression of the Effect of Temperature on Microalgae Productivity." *Biofizika* (USSR), 14, 196 (1969); *Biol. Abs.*, 50, 77558 (1969).
167. Smayda, T. J., "Experimental Observations on the Influence of Temperature, Light, and Salinity on Cell Division of the Marine Diatom, *Detonula confervacea* (Cleve) Gran." *Jour. Phycol.*, 5, 150 (1969).
168. Smirnova, L. I., "Physiology of Granular Leucocytes in Fish Blood." *Prob. in Ichthyology* (USSR), Amer. Fish Soc. Translation, 8, 748 (1968, 1969).
169. Snyder, G. R., "Effects of Heated Discharges on Anadromous Fishes: Discussion." In "Biological Aspects of Thermal Pollution." Krenkel, P. A., and Parker, F. L., Eds., Vanderbilt Univ. Press, Nashville, Tenn., 318 (1969).
170. Somero, G. N., "Enzymic Mechanisms of Temperature Compensation: Immediate and Evolutionary Effects of Temperature on Enzymes of Aquatic Poikilotherms." *Amer. Naturalist*, 103, 517 (1969).
171. Sopina, V. A., "Interclonal Differences in Thermostability in Amoeba." *Tsitologiya* (USSR), 10, 207 (1968); *Biol. Abs.*, 50, 134521 (1969).
172. Spitzer, K. W., Marvin, D. E., Jr., and Heath, A. G., "The Effect of Temperature on the Respiratory and Cardiac Response of the Bluegill Sunfish to Hypoxia." *Comp. Biochem. Physiol.*, 30, 83 (1969).
173. Stepanskii, V. I., and Yaglova, L. G., "Temperature Dependence of the Membrane Potential of *Nitella mucronata* Cells." *Biol. Nauk.* (USSR), 11, 79 (1968); *Biol. Abs.*, 50, 133591 (1969).
174. Stephenson, E. M. and Potter, I. C., "Temperature and Other Environmental Effects on Ammocoete Organs in Culture." *Jour. Embryol. Exp. Morphol.*, 17, 441 (1967); *Biol. Abs.*, 50, 87000 (1969).
175. Stewart, J. E., Cornick, J. W., and Zwicker, B. M., "Influence of Temperature on *Gaffkemia*, a Bacterial Disease of the Lobster *Homarus americanus*." *Jour. Fish. Res. Bd. Can.*, 26, 2503 (1969).
176. Stockner, J. G., "The Ecology of a Diatom Community in a Thermal Stream." *Brit. Phycol. Bull.*, 3, 501 (1968).
177. Sukhanova, K. M., "Temperature Adaptation in Protozoa." *Nauka* (USSR), (1968); *Biol. Abs.*, 50, 134522 (1969).
178. de Silva, D. P., "Theoretical Considerations of the Effects of Heated Effluents on Marine Fishes." In "Biological Aspects of Thermal Pollution." Krenkel, P. A., and Parker, F. L., Eds., Vanderbilt Univ. Press, Nashville, Tenn., 229 (1969).
179. Tagatz, M. E., "Some Relations of Temperature Acclimation and Salinity to Thermal Tolerance of the Blue Crab, *Callinectes sapidus*." *Trans. Amer. Fish. Soc.*, 98, 713 (1969).
180. Tageeva, S. V., and Kazantsev, E. N., "Statistical Distribution of the Velocity of Cytoplasm and Chloroplasts at Various Temperatures (*Elodea*)." *Fiziol. Rast.* (USSR), 15, 402 (1968); *Biol. Abs.*, 50, 78908 (1969).
181. Tatarko, K. I., "The Effect of Temperature on the Meristic Characters of Fishes." *Prob. in Ichthyol.* (USSR), *Amer. Fish. Soc. Translation*, 8, 339 (1968, 1969).
182. Tatarko, K. I., "Sensitivity of Pond Carp to Elevated Temperature at Different Periods of Embryonic Development." *Gidrobiol. Zh., Acad. Nauk. Ukr.* (USSR), 4, 34 (1968); *Biol. Abs.*, 50, 104456 (1969).
183. Tett, P. B., "The Effects of Temperature upon the Flash-Stimulated Luminescence of the Euphausid *Thysanoessa raschi*." *Jour. Mar. Biol. Ass. U. K.*, 49, 245 (1969).
184. Toews, D. P., and Hickman, C. P., Jr.,

"The Effect of Cycling Temperatures on Electrolyte Balance in Skeletal Muscle and Plasma of Rainbow Trout, *Salmo gairdneri*." *Comp. Biochem. Physiol.*, **29**, 905 (1969).

185. Toriumi, M., "Analysis of Intraspecific Variation in *Lophopodella carteri* (Hyatt) from the Taxonomic Viewpoint: XI. Preliminary Observations on the After Effect of Temperature on the Spine Number of the Spinoblast." *Sci. Rep. Tohoku Univ., Ser. IV, Biol.* (Japan), **33**, 487 (1967); *Biol. Abs.*, **50**, 128800 (1969).
186. Trpis, M., and Horsfall, W. R., "Development of *Aedes sticticus* (Meigen) in Relation to Temperature, Diet, Density and Depth." *Ann. Zool. Fenn.*, **6**, 156 (1969); *Biol. Abs.*, **50**, 117809 (1969).
187. Trpis, M., and Shemanchuk, J. A., "The Effect of Temperature on Pre-Adult Development of *Aedes flavescens* (Diptera: Culicidae)." *Can. Entomol.*, **101**, 128 (1969).
188. Trukhin, N. J., and Mikryakova, T. F., "Effect of Temperature on the Growth of *Chlorella* in Intensive Culture." *Fiziol. Rast.* (USSR), **16**, 432 (1969); *Biol. Abs.*, **50**, 128135 (1969).
189. U. S. Federal Power Commission, "Problems in Disposal of Waste Heat from Steam-Electric Plants." U. S. Govt. Printing Office, Washington, D. C., 50 pp. (1969).
190. van Uden, N., Abranches, P., and Cabeca-Silva, C., "Temperature Functions of Thermal Death in Yeasts and Their Relation to the Maximum Temperature for Growth." *Arch. Mikrobiol.*, **61**, 381 (1968).
191. Vasetskii, S. G., "Changes in the Ploidy of Sturgeon Larvae Induced by Thermal Treatment of the Eggs at Various Stages of Development." *Dokl. Akad. Nauk. SSSR* (USSR), **172**, 1234 (1967); *Biol. Abs.*, **50**, 87808 (1969).
192. Vernberg, W. B., and Vernberg, F. J., "Interrelationships Between Parasites and Their Hosts: IV. Cytochrome C Oxidase Thermal-Acclimation Patterns in a Larval Trematode and its Host." *Exptl. Parasitol.*, **23**, 347 (1968); *Biol. Abs.*, **50**, 55931 (1969).
193. Vincent, F., "Spawning Ecology of the White Bass *Roccus chrysops* (Rafinesque) in Utah Lake, Utah." *Great Basin Naturalist*, **28**, 63 (1968).
194. Webber, H. H., and Giese, A. C., "Reproductive Cycle and Gametogenesis in the Black Abalone *Haliotis cracheroidii* (Gastropoda: Prosobranchiata)." *Marine Biol.* (West Germany), **4**, 152 (1969).
195. Welch, E. B., "Ecological Changes of Applied Significance Induced by the Discharge of Heated Waters: Discussion." In "Engineering Aspects of Thermal Pollution." Parker, F. L., and Krenkel, P. A., Eds., Vanderbilt Univ. Press, Nashville, Tenn., 58 (1969).
196. Westin, L., "Lethal Limits of Temperature for Fourhorn Sculpin *Myoxocephalus quadricornis* (L.)." *Rep. Inst. Freshwat. Res. Drottningholm* (Sweden), **48**, 71 (1968).
197. Winter, J. E., "On the Influence of Food Concentration and Other Factors in Filtration and Food Utilization in the Mussels *Arctica islandica* and *Modiolus modiolus*." *Marine Biol.*, **4**, 87 (1969).
198. Winterbourn, M. J., "The Faunas of Thermal Waters in New Zealand." *Tuatara* (New Zealand), **16**, 111 (1968); *Biol. Abs.*, **50**, 107486 (1969).
199. Wurtz, C. B., "The Effects of Heated Discharges on Freshwater Benthos." In "Biological Aspects of Thermal Pollution." Krenkel, P. A., and Parker, F. L., Eds., Vanderbilt Univ. Press, Nashville, Tenn., 199 (1969).
200. Yanase, R., and Imai, T., "The Effect of Light Intensity and Temperature on the Growth of Several Marine Algae Useful for Rearing Molluscan Larvae." *Tohoku Jour. Agr. Res.* (Japan), **19**, 75 (1968); *Biol. Abs.*, **50**, 27841 (1969).

ECOLOGICAL MANAGEMENT PROBLEMS CAUSED BY HEATED WASTE WATER DISCHARGE INTO THE AQUATIC ENVIRONMENT

John Cairns, Jr.

The invitation to present a paper at this conference indicated that my assignment was to discuss physiological effects of heat upon fish (reproduction, metabolism, avoidance, attraction, etc.) and upon lower organisms. As I dutifully organized my thoughts along these lines I suddenly realized that this paper sounded very much like a trio of papers written nearly fourteen years ago (Cairns, 1955, 1956a, 1956b). Of course there is much more evidence now that increased temperature does affect aquatic organisms (Figure 1) and will produce adverse effects if the change is great or the organisms are already near their tolerance limit. Rather than retread this already well-beaten path it seemed best to turn to ecological management problems related to heated waste water discharge since these are the ones which need our attention most. Many of these have been brought to my attention by correspondence generated by a recent paper (Cairns, 1968).

```
1880--XX
1885--X
1887--X
1892--X
1894--X
1895--XXX
1897--X
1898--XXXX
1899--XXX
1901--X
1902--XXX
1903--X
1905--X
1906--XX
1908--X
1909--XXX
1910--X
1911--X
1912--XXXXX
1913--XX
1914--XXXXXXX
1915--XXX
1916--XX
1917--X
1918--XX
1919--XXXXX
1920--XXX
1921--XXXXX
1922--XXXXX
1923--XX
1924--XXXXXXX
1925--XXXXXX
1926--XXXXXXXXXX
1927--XXXXXXXXXXXXXX
1928--XXXXXXXXXXXXXX
1929--XXXXXXXXXXXXXXXXXXXXXXX
1930--XXXXXXXXXXXXX
1931--XXXXXXXXXXX
1932--XXXXXXXXXXXX
1933--XXXXXXXXXX
1934--XXXXXXXXXXXXXXXX
1935--XXXXXXXXXXXX
1936--XXXXXXXXXXXXXXXXX
1937--XXXXXXXXXXXXXXXX
1938--XXXXXXXXXXXXXX
1939--XXXXXXXXXXXXXXXXXXXX
1940--XXXXXXXXXXXXXXXXXX
1941--XXXXXXXXXXXXXX
1942--XXXXXXXXXXXXXXXX
1943--XXXXXXXXXXXXXX
1944--XXXXXXXXXXX
1945--XXXXXXXX
1946--XXXXXXXXXXXXXXXXX
1947--XXXXXXXXXXXXX
1948--XXXXXXXXXXXX
1949--XXXXXXXXXXXXXX
1950--XXXXXXXXXXXXXXXXXXXXX
1951--XXXXXXXXXXXXXXX
1952--XXXXXXXXXXXXXXXXXXX
1953--XXXXXXXXXXXXXXXXXXXXXXXXXXXX
1954--XXXXXXXXXXXXXXXXXXXXXXXXXXXXXXXX
1955--XXXXXXXXXXXXXXXXXXXXXXXXXXXXXXXXXX
1956--XXXXXXXXXXXXXXXXXXXXXXXXXXXXXXXXXXXXXXXX
1957--XXXXXXXXXXXXXXXXXXXXXXXXXXXXXXXXXXXXXXXXXXXXXXXXXXX
1958--XXXXXXXXXXXXXXXXXXXXXXXXXXXXXXXXXXXXXXXXXXXXXXXXXXXXXXX
1959--XXXXXXXXXXXXXXXXXXXXXXXXXXXXXXXXXXXXXXXXXXXXXXXXXXX
1960--XXXXXXXXXXXXXXXXXXXXXXXXXXXXXXXXXXXXXXXXXXXXXXXXXXXXXXX
1961--XXXXXXXXXXXXXXXXXXXXXXXXXXXXXXXXXXXXXXXXXXXXXXX
1962--XXXXXXXXXXXXXXXXXXXXXXXXXXXXXXXXXXXXXXXXXXXXXXXXXXXXXXXX
1963--XXXXXXXXXXXXXXXXXXXXXXXXXXXXXXXXXXXXXXXXXXXXXXXXXXXXXXX
1964--XXXXXXXXXXXXXXXXXXXXXXXXXXXXXXXXXXXXXXX
1965--XXXXXXXXXXXXXXXXXXXXXXXXXXXXXXXXXXXXXXXXXXXXXXXXXXXXX
1966--XXXXXXXXXXXXXXXXXXXXXXXXXXXXXXXXXXXXXXXXXXX
```

FIGURE 1. The distribution by year of literature on the effects of temperature in the aquatic environment. Data taken from Kennedy and Mihursky (1967).

At this conference we are probably going to talk about heated waste water discharge into the environment as if it were a problem that can be considered in isolation from other problems. There are good and valid reasons for doing this, the most important being that the heated waste water discharge problem alone is so complex that it is difficult for most people, even those who have worked in this field for years, to generalize about it. In fact it has been said that the ability to generalize is inversely related to one's knowledge about a problem. If this is true, perhaps meetings of this sort which force us to generalize are the only way to cope with today's complex problems. However, we should recognize that the response of an aquatic community to heated waste water will be determined in large part by other factors affecting the community. Have these aquatic organisms been exposed to insecticides, sewage, industrial wastes, heavy silt loads, and are they well nourished and vigorous or infected with various disease organisms? The response of an aquatic community will be determined by the aggregate stress conditions to which it is exposed, and the response will represent a summation of these stresses. Therefore if one asks if a specific temperature increase will affect fish or other aquatic organisms one must know not only the usual ecological requirements of these species but must also be able to estimate the total stress put upon these species by other forms of waste discharge (agricultural, industrial, municipal, etc.). In short, I am recommending that each drainage basin be regarded as a single operating unit (Cairns, 1967) and that the effect of all stresses upon the aquatic organisms be considered as well as the effect of individual stresses. This will not be easy! Our present laws and assessment techniques are designed primarily to cope with wastes individually rather than collectively. In addition, programs for systems analysis of an entire drainage basin are still rather primitive and inadequate although they could be developed quite rapidly (Cairns and Humphrey, 1969). However, we should recognize that environmental quality control and effective regional management will only be possible if we treat drainage basins as single systems and design waste loadings on that basis rather than one fragment at a time.

Having given my views on desirable future goals for water management I will turn to present-day problems with heated waste water discharges into the aquatic environment. Discharge of heated waste water seems to have the same effects as other types of stress in that there are three levels of biological response:

a range within which no abnormal response to change is noted;

a zone of graded response - increased stress produces increased response (more dead fish, fewer species, etc.);

a threshold beyond which further increases elicit no further response (everything is dead).

Most aquatic organisms can withstand the effects of heat only within very narrow limits. They are further handicapped by not having regulatory mechanisms (such as sweating) which buffer effects of increased heat in the environment. It is true that some can move to cooler areas, but this is avoidance rather than regulation and is possible only when a cooler area exists. Moreover, the discharge of heated water into a stream can create a hot water barrier which effectively blocks the spawning migrations of many species of fish, as well as other, less spectacular movements. The temperature an aquatic animal can tolerate depends greatly upon the conditions in the normal environment--tropical fishes are adapted to warm waters that would kill a brook trout. Sometimes in areas with very stable temperatures the margin of safety is small. The polyps that build coral reefs are killed by temperatures only two or three degrees above those at which they carry out a normal existence.

Though abrupt application of heat can kill bacteria and other microorganisms (pasteurization is an example) the probable result of gradual changes in a natural environment is exclusion of non-tolerant species by resistant species. That is, those species more tolerant of the new conditions outperform the others and reproduce more rapidly while the less tolerant reproduce slowly or not at all. The result is a qualitative shift in kinds of species present which may or may not be accompanied by a change in total numbers of individuals (Cairns, 1956a). One group of algae called *diatoms*, another called *greens*, and a third called *blue-greens*, each group including a number of species, may all be found together in an aquatic ecosystem. However, Figure 2 shows diagrammatically how the species relationships shift as the temperature rises from 68°F (20° C) to 104°F (40° C). The number of species of diatoms drops sharply; the trend among blue-greens is the opposite--there are more species present in the warmer water. Species of greens also increase with rising temperature until it passes 86°F (30° C) then they drop off again. Present evidence suggests that blue-greens are not as suitable as diatoms to the organisms dependent on algae as a food supply. Many blue-greens produce unpleasant odors and some are toxic to shellfish.

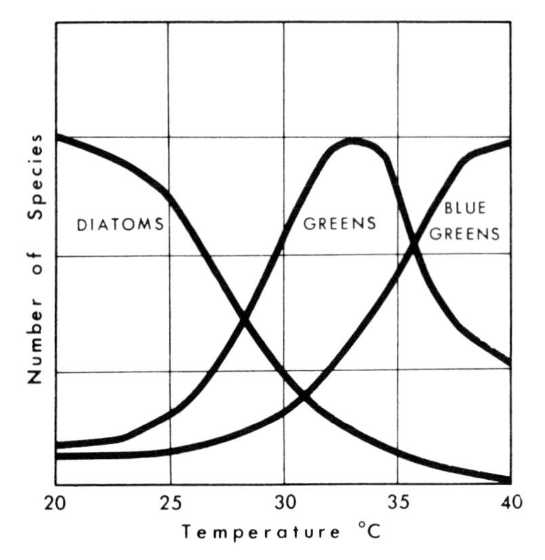

FIGURE 2. Qualitative shifts in algal population composition with temperature (from Cairns, 1956a).

The major effects of heat pollution upon higher aquatic organisms are:

1. Death through direct effects of heat.
2. Internal functional aberrations (changes in respiration, growth).
3. Death through indirect effects of heat (reduced oxygen, disruption of food supply, decreased resistance to toxic substances).
4. Interference with spawning or other critical activities in the life cycle.
5. Competitive replacement by more tolerant species as a result of the above physiological effects.

Experiments suggest that some aquatic organisms can become acclimated to higher temperatures than those to which they are normally accustomed. But even in the laboratory, gradual acclimation to much higher temperatures has limitations. The experiments of Alabaster (1963) have shown that for three species of fish, the mean lethal *rise* in temperature *decreases* as acclimatization temperature increases

(Figure 3). The significance of these results is that the higher the temperature to which one exposes fish, *even within their tolerance limits*, the more closely one must control temperature changes. This means increased management costs! Of course, one should be quite cautious about using results even from carefully controlled laboratory experiments to develop standards for use in natural environments. In the laboratory, an organism may have near optimal conditions with the exception of the changing temperature being studied. Many organisms in the natural environment already live under less than optimal conditions and are often subject to other kinds of stress at the same time the temperature is changing. In fact, another form of stress is a secondary effect of heated water. Heating water decreases the solubility of oxygen (Figure 4) and, within certain limits, increases bacterial activity. These events lower the amount of oxygen available to the higher aquatic organisms.

Initial attempts to reduce biological problems resulting from the discharge of heated waste waters were primarily directed toward avoiding lethal temperatures. However, since temperature changes often regulate the occurrence of biological events, it is quite possible to disrupt biological systems fundamentally while staying within non-lethal temperatures by changing the *pattern* of temperature change. Gardeners know that certain bulbs and seeds require *cooling* in order to produce plants. The temperature pattern during the cooler months also affects the life cycles of many aquatic organisms so that where heated water is discharged the receiving water *may not get cold enough* for many species. In addition the emergence pattern and reproductive cycle of some aquatic insects is dependent upon seasonal temperature cycles. A population of insects emerging in March when normally it would appear in May would surely meet with disaster. Further, its absence in May would be likely to affect other animals which depend on it for food. A whole chain of events that would seriously disrupt normal biological interactions could be set into motion by the mistiming of seasonal temperature changes.[3]

These few examples give an indication of some of the ways in which increased temperatures from heated waste water discharge can affect an aquatic community. Those wishing further information can consult the enormous volume of references recently produced by Raney and Menzel (1969). Other worthwhile source materials can be found in Kennedy and Mihursky (1967), Raney and Menzel (1967), Prosser (1967), and Holdaway *et al.* (1967). The rapid increase in literature on the effects of temperature changes upon aquatic organisms is illustrated in Figure 4. Obviously, though we lack information which enables us to predict the *precise response* of aquatic organisms in a *particular* stream, lake, or area of the ocean with precision, we have an enormous amount of basic literature most of which indicates that organisms can usually tolerate small changes in the temperature of their environment, that further increases will probably produce problems, and that substantial increases will produce serious problems. Our most pressing need is not to further document the effects of increased temperature upon aquatic organisms but rather to develop an operational approach which will enable us to deal with these problems rationally and effectively.

The related and complementary problems of increased energy production and human population growth are forcing us to make a choice between a complex environment with considerable functional stability and a simplified environment with *increased management costs*. A good example of a complex environment is a forest consisting of a great variety of plants and animals which will persist year in and year out with no interference from man. This ecosystem is a complex mixture of biological, chemical, and physical interactions, many of which cancel each other out. For example, if there are several predators regulating a population of rodents and one disappears, the effect is often reduced by a population expansion of other predators also feeding upon the rodents. Therefore, the system is one of *dynamic equilibrium*, with the system itself being stable but with many of its components undergoing change. A

---

[3]This explanation is necessarily oversimplified. For readers who would like to learn more about ecosystems, I recommend *Ecology* by Eugene P. Odum, published by Holt, Rinehart and Winston, New York, 1963.

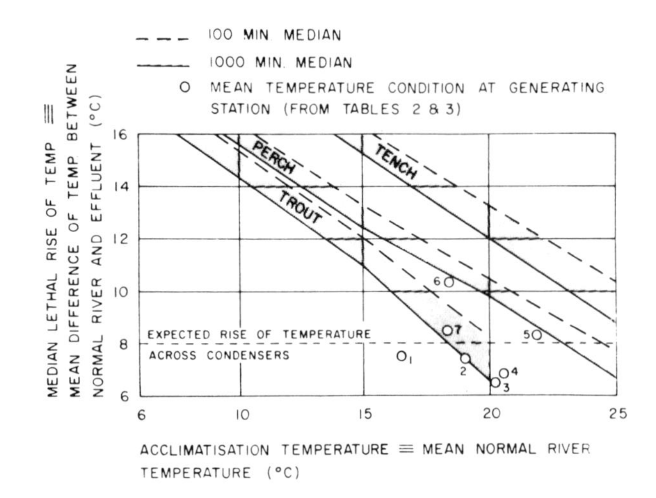

FIGURE 3. For three species of fish the mean lethal *rise* in temperature *decreases* as acclimatization temperature increases (from Alabaster, 1963).

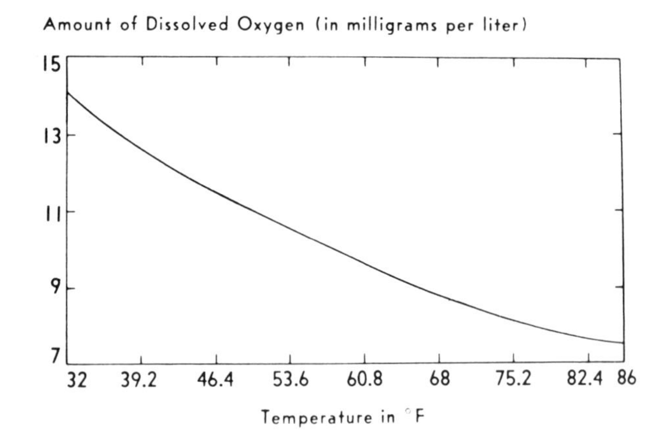

FIGURE 4. Reduction in solubility of dissolved oxygen as temperature increases.

simple system such as a corn field produces a quantity of material immediately useful to man, but is notoriously unstable. Without constant care and attention it would cease to be useful and would disappear. So the history of civilization has been one of widespread simplification of the environment with consequent *increased management requirements*.

Kenneth Boulding (1966) has suggested that we are now in a transitional state between a "cowboy" economy and a "spaceship" economy. The essence of his ideas is that we are moving from a frontier psychology in which we solved our problems by moving to a new area when we had destroyed the old one or depleted its resources to a spaceship psychology in which we are becoming aware that we are living on a non-expandable planet with finite resources. In short, what we formerly regarded as an open system we now are beginning to regard as a closed system. The transition will be painful since the happy-go-lucky frontier days in which there was a surplus of almost everything if one had the energy to get it are much more appealing than "spaceship life" in which there must be careful and precise utilization of everything and particularly in which limitations and restrictions on use are the rule rather than the exception. Nevertheless, with our rapidly increasing population and finite resources that is where we are today!

Our basic problem is that we are dependent upon a life-support system which is partly industrial and partly ecological. Unfortunately, we have reached a stage of development where the non-expandable, ecological portion of our life-support system is endangered by the expanding industrial portion. Optimal function and full beneficial use of both portions of our life-support system will only be possible if a variety of disciplines and diverse points of view can cooperate and work together effectively.

There are four basic alternatives which are open to us with regard to the heated waste water problem which we may choose singly or in various combinations:

(1) Placing all heated waste water in streams, lakes, and oceans without regard to the effect, and considering the environmental damage as a necessary consequence of our increased power demand.

(2) Using, but not abusing, present ecosystems. This means regulating the heated waste water discharge to fit the receiving capacity of the ecosystem.

(3) Finding alternative ways to dissipate or beneficially use waste heat.

(4) Modifying ecosystems to fit the new temperature conditions.

The first choice will obviously lead to ecological disaster but unfortunately is attractive to some people. The second, using but not abusing ecosystems, would entail *determining the receiving capacity of each ecosystem for all the wastes being discharged into it*. This would include insecticide run-off, sewage, silt from road building, and heavy metals, as well as heated waste waters. This alternative will probably not please either conservationists or industrialists, but it will partly satisfy the hopes of each group and should be an acceptable compromise. With an expanding population and finite resources it is not likely that either group can have many sizeable areas set aside for exclusive use. We should of course protect our National and State Parks and other remaining natural areas. In the rest of the country multiple use will undoubtedly become increasingly necessary. The rapidly growing economic importance of recreation should not be overlooked when multiple use plans are developed.

The third alternative--that of finding other ways of dissipating heat or using it for beneficial purposes (heating homes, hydroponic units in the northern states in winter, etc.)--is one that deserves much more attention than it has received in the past. Industry should take the initiative here.

The fourth alternative appears very attractive initially (Cairns, 1969) since we are accustomed to modifying the environment to suit our own immediate needs. Furthermore since Naylor (1965) and others have shown that warmer water species may become established in heated waste water discharge areas accidently, why not do this deliberately? The catch is threefold: (1) the management costs would increase since people experienced in aquaculture would be required to introduce and manage the new species; (2) there would be lag time while the warmer water species replaced the colder water species during which the area would not be ecologically functional; and most importantly (3) the new species would be *dependent* on the heated waste water discharge and might die if it were shut off due to a strike or a temporary cessation of operations due to other causes. The latter event should raise some interesting legal issues since the death of the organisms would be due to *withholding* heat upon which the aquatic community had become dependent.

There is no question that it is possible to discharge controlled amounts of heated water into the aquatic environment without damaging it for other beneficial uses, or even without significantly degrading the aquatic community. This has been shown by Cairns (1966a), Patrick, Cairns, and Roback (1967), Cairns (in press, a), Fetterolf and Oeming (in press), and others as well. The problem we face is not whether it is possible to discharge heated waste water into the aquatic environment without seriously degrading it, but rather to determine the receiving capacity of each aquatic environment which will undoubtedly have unique characteristics which will effect the amount of heated waste water that can safely be introduced. It is my belief that fixed arbitrary regulatory standards which do not consider the unique ecological characteristics of a particular area may not permit full utilization of the receiving capacity of the aquatic environment for heated waste water in some cases, or may not fully protect the aquatic environment from the heated waste water discharge in others.

Since I have already indicated that reaching a harmonious relationship with our environment will depend more on management practices than on generating more scientific information I feel a few suggestions for a future approach toward the problem of environmental management are in order. If we agree that we are living in a closed system, "spaceship earth," in which we must view the environment as a life-support system for an expanding human population our goals can be defined in significantly different terms. Instead of trying to determine how much environmental degradation, if any, is acceptable to us we might start by asking what functions of the environment as a life-support system are essential to our survival? It is quite obvious that we need ecosystems to produce oxygen, degrade and transform waste products, produce food, and for a variety of other beneficial purposes. We have taken our ecosystems for granted in the past because "spaceship earth" was not crowded and pressures for understanding its life-support system were not great. Since the system was working well without any attention management practices designed to protect and enhance it were not developed. However, with a rapidly increasing population and evidences of serious environmental "backlashes" all around us the functioning of our life-support system has become an urgent problem which had better receive our immediate attention.

As a first step I would suggest that industry and other waste dischargers (including municipalities and agricultural groups) be brought more fully into environmental management problems and be made responsible for environmental quality control in the receiving stream or other body of water to a much greater extent than is now the case. (However it is only fair to note that many industries have had effective environmental quality control programs for years.) Although it is obviously impossible to outline a procedural method for all situations, a basic plan might include:

1. Using an ecologist in the site selection process. Much trouble could be avoided if ecological considerations were included at this stage.
2. Carrying out a pre-operational ecological survey which includes biological, chemical, and physical information to get a precise indication of ecological conditions before operations begin.

3. Carrying out some laboratory experiments simulating both environmental conditions and prospective waste discharges to determine the effects upon representative species common to the discharge area.

4. Developing a series of waste treatment and disposal practices based on the above data.

5. Utilization of continually operative biological monitoring systems both "in stream" and "in plant" (some of these are described in Cairns (1966, 1967) and Cairns *et al.* (1970). These monitoring systems, particularly the "in stream" units, should be in operation before discharge begins.

6. One or more post-operative surveys to determine the effectiveness of the estimates based on the simulated discharge experiments and to determine whether any degradation is occurring in the receiving body of water.

7. Continued alteration of operating procedures to bring waste discharges into line with the carrying capacity of the system. (As each ecosystem approaches its *maximum* carrying capacity adding a new industry or other waste discharge to the system will require reduction of other waste discharges. The same will be true for increased waste discharge from an existing industry.)

It is my belief that any industry willing to furnish sound scientific evidence that its waste discharges are not degrading the receiving stream should be permitted to make full use of its receiving capacity. The corollary to this is that if the waste discharge significantly degrades the ecosystem the situation should be corrected immediately even if the waste discharge conforms to state standards.

The above recommendations are, of course, for a single waste discharger although it is quite obvious that the discharge of one industry, municipality, or agricultural run-off will probably affect the receiving capacity of the stream for other wastes. Therefore it is also obvious that it is in the best interests of at least the majority of users of the receiving capacity of an ecosystem to cooperate so that maximum beneficial use may be made of the ecosystem by all users. If one industry or municipality is using practically all of the receiving capacity of a stream there may be some reluctance to cooperate with other groups who will be the primary beneficiaries of increased treatment. However, as the space per person decreases and the utilization of this space increases it is in our own enlightened self-interest to cooperate in this fashion. When this is done, the systems approach toward an entire drainage basin or other ecological unit becomes feasible. The systems approach might be divided into three basic components (modified from Watt, 1966):

1. Systems analysis - determining the operational characteristics of a system (i.e., determining the cause-effect pathways, etc.).

2. Systems simulation - simulating the various alternative uses which may be made of a system.

3. Systems optimization - selecting that combination of uses which provide the greatest benefits to the greatest number of people.

Some of the techniques for the systems approach already developed by the Army Corps of Engineers could be modified to water quality management problems (Cairns and Humphrey, 1969). In any case our problems are too complex and too interdependent to be solved one at a time. Past experience has clearly shown that our ecosystems and our whole society are too interwoven to treat any portion of their operations in isolation.

In summary, almost all of the environmental problems we face today are a function of our increasing population, our increasing individual expectations, and the finite resources we have to support these demands. We could successfully treat problems in isolation in the past because there was sufficient space for them to be truly isolated. As our utilization of existing space increased, more and more of these problems began to be interrelated. As the number of complex relationships increases we are forced to turn away from fixed standards which were developed to handle isolated problems, to developing management practices and techniques designed to resolve complex, interrelated problems. This is as true in the waste disposal field as it is in any other field today. Therefore my plea is to forego attempts to establish fixed numerical standards for discharge of heated waste water or any other waste material and to turn instead to determining those qualities of an ecosystem that are important to us and basing utilization of the ecosystem upon maintaining these. This will mean developing a managerial group for ecosystem management which is responsible for maintaining quality and insuring proper function. No doubt the transitional period will have some unpleasant features as do all transitional periods, but it is quite obvious that our old view of our relationship to the environment is hopelessly outdated. This is one of the reasons why we cannot develop standards for heated waste water discharge which are acceptable to regulatory agencies, industries, legislators, and other interested groups. This failure will persist until our basic goals are redefined and management of the environment in the best sense of the word becomes a reality.

ACKNOWLEDGEMENTS

The writer appreciates the comments of Professor Arlen Edgar, Professor Arthur Cooper, Miss Jeanne Ruthven, and Mr. William Yongue during the preparation of this manuscript.

LITERATURE CITED

Alabaster, J. S. 1963. Effects of heated effluents on fish. Jour. Air and Water Poll. 7(6/7):541-563.

Boulding, Kenneth E. 1966. The economics of the coming spaceship earth. Reprinted in Population, Evolution, and Birth Control, edited by Garret Hardin, pp. 115-119, published in April, 1969, by W. H. Freeman and Company. The original article was in Environmental Quality in a Growing Economy, Henry Jarret, Editor, The Johns Hopkins Press, Resources for the Future, Inc.

Cairns, John, Jr. 1955. The effect of increased temperature upon aquatic organisms. Proc. 10th Industrial Waste Conf., Purdue Univ., Engineering Bulletin, No. 89, pp. 346-354.

_______. 1956a. Effects of increased temperatures on aquatic organisms. Indus. Wastes 1(4):150-152.

_______. 1956b. Effects of heat on fish. Indus. Wastes 1(5):180-183.

_______. 1966a. Biological concepts and industrial waste disposal problems. Proc. 20th Indus. Waste Conf., Ext. Series No. 118, Purdue University, pp. 49-59.

_______. 1966b. The protozoa of the Potomac River from Point of Rocks to Whites Ferry. Not. Nat. Acad. Nat. Sci. Phila., No. 387, pp. 1-11 plus 43 pp. support-data deposited as document No. 8902 with the AID Aux. Pub. Proj. Photodupl. Serv., Library of Congress, microfilm copies $2.50.

_______. 1967. The use of quality control techniques in the management of aquatic ecosystems. Water Resources Bull. 3(4):47-53.

_______. 1968. We're in hot water! Scientist and Citizen 10(8):187-198.

_______, and P. S. Humphrey. 1969. A water resources ecology capability for the

Waterways Experiment Station and the U.S. Army Corps of Engineers. U.S. Army Corps of Engineers Contract Report 0-69-1, 26 pp.

_______. 1969. Formal discussion of "The effects of heated discharges on fresh-water benthos" by Charles B. Wurtz. Biological Aspects of Thermal Pollution, Vanderbilt University Press, pp. 214-220.

_______. (In Press, a). The response of aquatic communities to thermal pollution and other forms of environmental stress. Farvar, M. Taghi and Milton, John P. (editors), Ecology and International Development. New York, Natural History Press, 1970.

_______, K. L. Dickson, R. E. Sparks, and W. T. Waller. 1970. Rapid biological information systems for water pollution control. Jour. Water Poll. Control Fed. 42(5), Pt. II, 685-703.

Fetterolf, Carlos M. and Loring F. Oeming. (In Press.) Formal discussion of "Water quality standards for temperature" by Robert S. Burd. National Symposium on Thermal Pollution: Engineering and Economic Consideration, Vanderbilt University Press.

Holdaway, J. E. *et al.* 1967. Temperature and Aquatic Life. Technical Advisory and Investigations Branch, Technical Services Program, Federal Water Pollution Control Administration, U.S. Department of Interior, Lab Investigation Series, No. 6.

Kennedy, V. S. and J. A. Mihursky. 1967. Bibliography on the effects of temperature in the aquatic environment. Natural Resources Institute, University of Maryland, mimeographed.

Naylor, E. 1965. Biological effect of a heated effluent in docks at Swansea, South Wales, Proceedings of the Zoological Society of London 114:253-268.

Patrick, R., J. Cairns, Jr. and S. S. Roback. 1967. An ecosystematic study of the fauna and flora of the Savannah River. Proc. Acad. Nat. Sci. Phila. 118(5): 109-407.

Prosser, C. Ladd. 1967. Molecular Mechanisms of Temperature Adaptation. American Association for the Advancement of Science, Pub. No. 84.

Raney, E. C. and B. W. Menzel. 1967. A Bibliography: Heated Effluents and Effects on Aquatic Life with Emphasis on Fishes. Cornell University, Ithaca, New York, Fernow Hall.

Raney, E. C. and B. W. Menzel. 1969. Heated Effluents and Effects on Aquatic Life with Emphasis on Fishes: A Bibliography. Cornell University Water Resources and Marine Sci. Center, Philadelphia Electric Co., and Ichthyological Associates, Bull. No. 2.

Watt, Kenneth E. F. 1966. Systems Analysis in Ecology. Academic Press, New York. 276 pp.

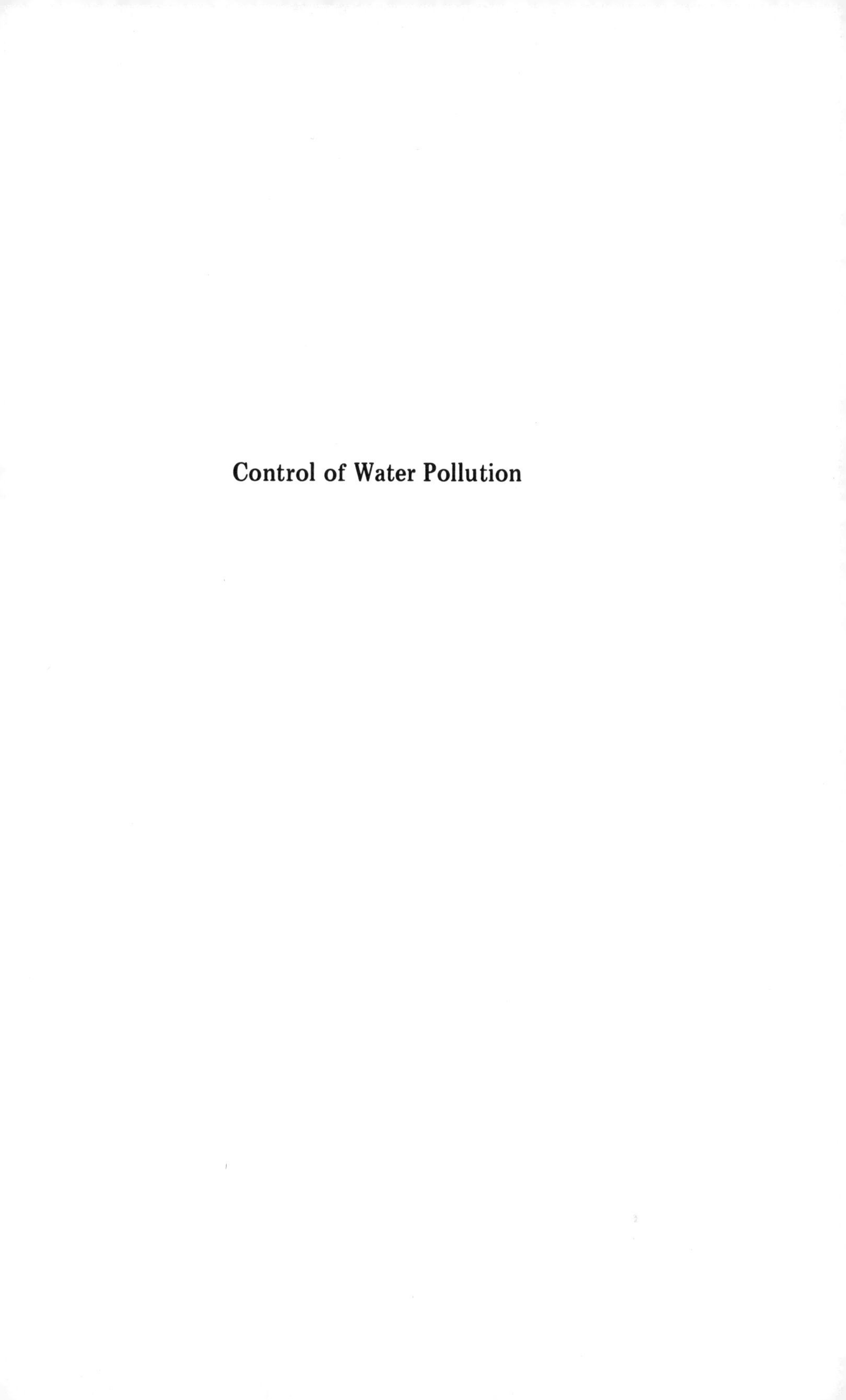

Control of Water Pollution

USE AND ABUSE OF NATURAL WATER SYSTEMS

A. W. Busch

The intent of this paper is to consider how *use* of natural water systems is sometimes *abuse*. Two major factors in abuse are intentional pollution for maximum use of stream assimilative capacity and the promotion of regional treatment systems for combined municipal-industrial wastes or for mixed industrial wastes.

Maximizing Pollution

The historical use of natural water systems to their maximum capability for disposal of man-made pollution is no longer a valid or viable engineering policy. As long as the amount of pollution "available" for discharge to a natural system did not exceed the amount that would significantly reduce quality parameters such as dissolved oxygen (DO) in the stream, intentional discharge of man-made pollution may have been justifiable. However, in considering partial treatment to avoid degrading a stream below some man-determined level, the policy becomes one of maximizing pollutional discharges, although disguised under the concept of fully utilizing stream assimilative capacity. In any growing society, and particularly a technologically oriented industrial society, this policy requires a continual reallocation, in incremental form, of the assimilative capacity of a given stream. Inasmuch as there will always be unintentional or uncontrollable pollution from such sources as stormwater runoff, agricultural drainage, spills, and illicit dumps, it would seem much simpler to require that there be no intentional discharge of pollutants based on the assimilative capacity of a stream.

The idea of using assimilative capacity dates back to the work of Streeter and Phelps who were able to define the sag curve of DO depression resulting from discharge of organic material to a natural system. This purely descriptive formulation has, since that time, been supported by a variety of workers as being of potentially predictive capability, particularly if augmented by a variety of increasingly esoteric mathematical models.

The basis most often quoted for use of the assimilative capacity of a stream is economics. It should be clear that what most people mean when they say that it is more economic to discharge pollution into a stream than to remove it in a treatment plant is that it costs less initially. A true economic analysis will involve the long-term accumulative impact of intentional discharge of pollution to limited natural systems. The present environmental crisis proves this point. In other words, long ago the solution to pollution was truly dilution. However, the dilution that would enable continuing to accept what was once a valid approach to waste disposal no longer exists. The consequence of following this policy for too long is the present estimate of billions of dollars for cleanup. Was the policy ever economic?

A major factor in the concept of

full use of stream assimilative capacity and attempts to use the Streeter-Phelps approach as a predictive methodology was the lack, for many years, of parameters other than biochemical oxygen demand (BOD). One of the more interesting aspects of the 5-day BOD test is that it is simultaneously sensitive and inaccurate. It is sensitive because it is suitable for dilute natural systems with concentrations of 10 mg/l as the norm. However, it is a singularly inaccurate test when applied to measurement of organic carbon. These seemingly contradictory statements may be used as the basis of defining one of the major difficulties in the field of pollution abatement. One must deal with highly dilute systems and must in effect magnify a pinhead so that its curvature can be measured.

None of the expensive studies carried out to define maximum stream assimilative capacity has ever recommended that the critical case is the *minimum* assimilative capacity and that this can be calculated in a straight-forward fashion. In other words, maximum assimilative capacity must involve probabilistic analysis and must establish confidence limits. Minimum assimilative capacity is defined easily and represents a firm quantity that cannot be exceeded without a gamble being involved. The minimum assimilative capacity of any body of water is its oxygen transfer capability under quiescent, summer temperature conditions. The interfacial oxygen transfer coefficient pertinent to this condition can be measured in a beaker in the laboratory. Typical K_2 values are also in the literature. For example, the minimum and maximum assimilative capacity of the Houston Ship Channel are on the order of 35,000 and 350,000 lb O_2/day (15,900 and 159,000 kg O_2/day) equivalents, respectively, at a deficit of 4 mg/l. The estimated daily discharge of oxygen equivalents to the Channel is 350,000 to 490,000 lb/day (159,000 to 222,000 kg/day). Not surprisingly, the Channel is partially anaerobic.

Perhaps intentional pollution can be put into proper perspective by asking, "Why not treat a portion of a waste completely and bypass the remainder instead of partially treating all of the waste?" This is likely to cost less.

Regional Treatment Systems

Another factor in use and abuse of natural water systems is widely held support for the concept of regional treatment plants as, again, more economic. To a large degree this concept is one of the inheritances from municipal waste treatment. Certainly there is no doubt that the same number of operators are sufficient for either a 1-mgd (3.785-cu m/day) or a 100-mgd (378,500-cu m/day) treatment plant. That is, the volumetric flow rate has little to do with the operational effort required. There is equally little doubt that the amount of concrete required per unit of tank volume is less for a large tank than for a small one. However, neither of these factors is of primary importance when considering the question of regional plants versus individual treatment plants for industrial wastes, for combined municipal and industrial wastes, or even, in some cases, for strictly municipal wastes. There is, of course, often a valid case for combining small municipal plants that would otherwise be unsupervised.

These concepts are all based on traditional design practices. An example of traditional design is the tendency always to equalize industrial wastes and never to equalize municipal wastewater. Another example is sedimentation basins that are rated on suspended solids (SS) removal but are designed functionally only for removal of settleable solids.

A basic problem is the lack of differentiation in the functions of the various treatment units. The parameters used to characterize wastewater fail to distinguish between the dissolved, colloidal, and suspended phases.

Another justification often used, or widely believed characteristic of regional plants, is that of more "efficient" treatment. A rather brief consideration of standard design procedure indicates a basic reason for the apparently higher "efficiency" of a large municipal treatment plant compared to a smaller one is that the ratio of peak to minimum flow is much greater for a small plant than it is for a large one. That is to say, small plants are basically closer to the source of wastewater than are large plants. This means there is less potential for volumetric damping in the interceptor sewer feeding a small treatment plant.

The regional plant concept for joint municipal-industrial treatment so widely held and so recently espoused by the Gulf Coast Waste Disposal Authority is suspect. A number of factors mitigate against joint municipal-industrial treatment. In many cases proper process design is overlooked or is not carried out because systems selection is made on a predetermined basis. As an example the pulp and paper industry, as represented by its National Council for Stream Improvement, seems to feel that 5 to 7 days of detention is a minimum for treatment of kraft mill waste. Two kraft mill biological waste treatment systems with a detention time in aeration of 3 hr are located in Texas. The justification for long detention time versus the actually required short detention time is "real." That is, if a Warburg oxygen uptake measurement is used for process design, then, of course, a long time is required for a set percentage of cumulative oxygen uptake to be obtained. However, carbon removal, that is, conversion of dissolved organic carbon into bacteria, has occurred very early in the reaction, normally in 2 to 3 hr, and the remainder of the oxygen uptake is caused by autoxidation of the cells produced from the carbon conversion.

As direct industrial water reuse becomes more obviously required, the use of regional plants is a disadvantage because the wastewater that is to be reused has been transported a substantial distance from the point where it is needed for reuse.

Several other primary faults with regional treatment systems are:

1. Reaction rate differences between various industrial wastes and between industrial and municipal wastes mean that the contributor of a rapidly degradable waste must pay for a disproportionate amount of residence time. Furthermore, disparities in organic removal rates frequently result in finely divided bacterial cell residue difficult to settle at conventional clarification rates.

2. For new combined systems financed by public bonds and built to serve industry, industrial contributors must sign long-term contracts commensurate with revenue bond terms, usually 20 yr or more.

3. Regional systems served by gravity sewers must accommodate a disproportionately greater amount of infiltration because of longer collection lines.

4. Regional systems built to replace existing municipal plants having outstanding bond issues must be supported by new taxes while taxpayers continue to pay for abandoned facilities.

5. Most municipal wastes do not require biological treatment because the collection system is a self-seeded, plug-flow reactor, and soluble carbon conversion usually has been accomplished in the sewer. The longer the residence time in the sewer the more likely this conversion is to be complete. Therefore, regional municipal systems are less likely to require a biological process than are small plants.

6. A mass balance will show, within the constraints of DO solubility, that a point source discharge of high volume must be of much higher quality than the same volume discharged at numerous points if stream quality is not to be degraded. A comparison is

a beam that can carry a much higher load uniformly distributed rather than concentrated at a point.

Justifying Intentional Pollution

A strong contrast exists between forecasts of population growth and continued advocation of maximum utilization of stream resources. Concern about how the various professionals involved in the field can justify intentional pollution is valid. For example, regulatory authorities are faced with periodic upgrading of effluent requirements if they fall into the trap of intentional pollution. Upgrading requires time and requires political decisions. The time required for upgrading may well present problems in water quality when multiple use of water is involved as in most river basins. How can the industrialist justify intentional pollution if subsequent and inevitable upgrading of effluent quality will make an installation unprofitable? How can the taxpayer justify intentional pollution when the additional cost of complete treatment is usually so small? How can the design engineer justify intentional pollution when the treatment reaction must be intentionally stopped at less than completion?

Summary

Use of natural water systems for intentional disposal of man-made waste is actually an abuse of natural systems and, even though validly justified 100 yr ago, is a concept that can no longer be tolerated. Instead, engineers should repudiate the concept of intentional pollution and partial treatment and get on with the business of abating pollution by applying the same stringent quality control approach to waste treatment as they do to manufacturing operations.

A five-minute solution for stream assimilative capacity

A. W. BUSCH

THE CAPACITY of natural waters to assimilate organic wastes is called the stream assimilative or purification capacity.[1] The intent of this paper is to show that the assimilative capacity that does not lower the oxygen content of the stream below a predetermined value is the significant capacity and should be the maximum capacity made available for intentional waste assimilation. Discharge of any wastes in excess of this minimum capacity must involve a probabilistic analysis. The minimum assimilative capacity is set by the minimum reaeration capacity because reaeration is the only continuous source of oxygen supply. The minimum reaeration capacity thus is the maximum usable assimilative capacity for uniformly applied and distributed wastes.

CONCEPTS

A previous paper[2] on aeration and oxygen transfer has pointed out that in waste treatment aeration the effectiveness of external energy input is constrained by the biological reaction. In other words, oxygen cannot be transferred faster than it is used. Natural reaeration of streams is exactly the opposite. In this case oxygen cannot be used unless it is transferred into the water by natural phenomena or unless instream aeration is practiced (Figure 1).

This means that the assimilative or purification capacity of a body of water is determined by the product of the minimum surface (mass) transfer coefficient, the maximum dissolved oxygen (DO) deficit, and the surface area being considered.

The general expression for gas transfer to a liquid is

$$\frac{dM}{dt} = K_L A (C_s - C) \qquad (1)$$

where

K_L = mass transfer coefficient,
A = interfacial surface area,
C_s = saturation DO concentration, and
C = ambient DO concentration.

For the worst condition in a natural body of water, that is, no surface turbulence and high temperature,

$$\frac{dM}{dt} = K_{L(\min)} D_{(\max)} A \qquad (2)$$

where

$D_{(\max)}$ = the maximum allowable DO deficit, and
A = the surface area of the stream or lake.

Equation 2 expresses the maximum amount of oxygen that can be transferred under the worst conditions. This is also the maximum amount of oxygen demand that can be uniformly imposed on a body of water without the DO concentration's dropping below that specified by the maximum deficit. A point source discharge must be of lesser quantity because the surface area available for reaeration is smaller.

There is a difference between predicting reaeration coefficients and establishing the assimilative capacity of a stream, lake, or estuary for assignment of waste discharge permits. Unfortunately, the literature often implies that the ability to predict reaeration

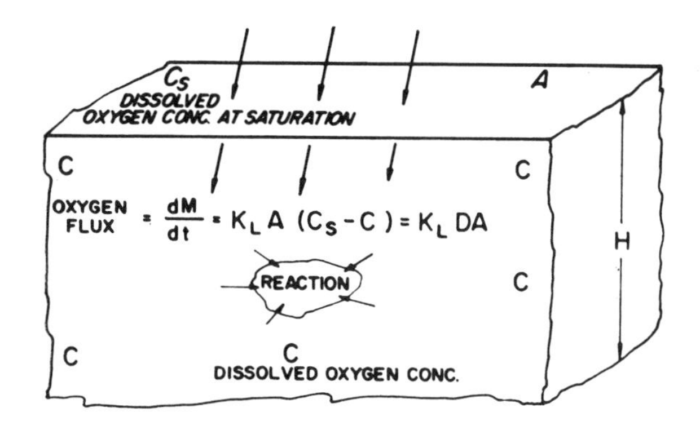

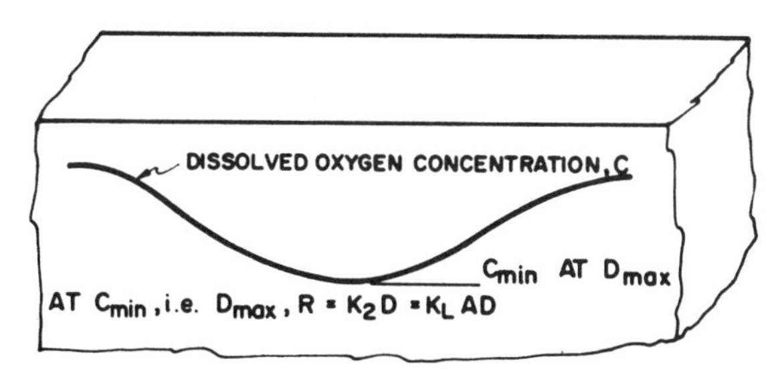

FIGURE 1.—Interfacial gas transfer (reaeration) in natural water.

coefficients is related to prediction of stream assimilative capacity. This is true only if a probability analysis is included in the predictive methodology. The average condition of a stream has no bearing on its minimum assimilative capacity, which depends on the instantaneous situation and is the maximum capacity available for intentional pollution.

Field Data

Several researchers have stressed the importance of surface conditions in reaeration. Churchill *et al.*[3] and Juliano[4] presented some of the best data showing the range of mass transfer capacity in natural waters. Their results can be used to project probable minimum transfer coefficients contrasted with the absolute minimum measured in the laboratory under quiescent conditions.

Churchill *et al.*[3] reported on studies of five rivers. K_L values calculated from their mean K_2 coefficients range from 1.9 to 9.1 ft/day (0.579 to 2.774 m/day). Juliano[4] reported minimum, average, and maximum K_2 values for seven stations on four rivers. Table I shows a summary of Juliano's[4] data and presents K_L values calculated from it. A minimum K_L of approximately

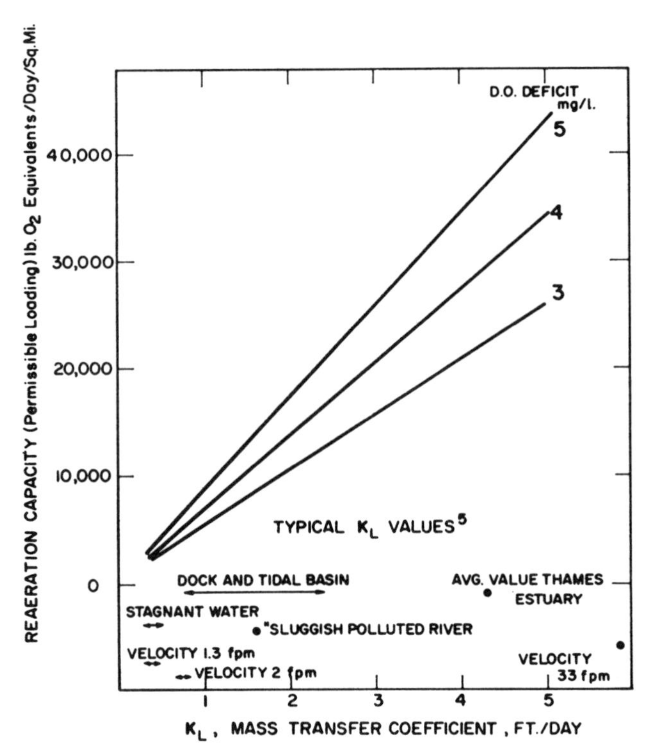

FIGURE 2.—Relationship of mass transfer coefficients, oxygen deficits, and reaeration capacity. (Fpm × 0.3048 = m/min; lb/day/sq miles × 0.175 = kg/day/sq km.)

2 is shown. Older literature contains K_L data ranging as low as 0.3 for stagnant water.

Using a minimum K_L of 2 ft/day (0.610 m/day) and a maximum deficit of 4 mg/l DO, Equation 2 shows that the maximum uniform waste load for a stream with these specifications is an oxygen equivalent of 13,800 lb/day/sq mile (2,419 kg/day/sq km) of water surface. A minimum K_L of 1 ft/day (0.305 m/day) for the same 4-mg/l DO deficit means a maximum oxygen demand of less than 7,000 lb/day/sq mile (1,227 kg/day/sq km). If $K_{L(min)}$ is 0.5 ft/day (0.152 m/day), the permissible demand is less than 3,500 lb/day/sq mile

TABLE I.—Reaeration and Mass Transfer Coefficients

Station*	Average Water, H* (ft)	Reaeration Coefficient K_2* (l/day)			Mass Transfer Coefficient K_L† (ft/day)		
		Max	Min	Avg	Max	Min	Avg
1	4	3.77	0.57	1.34	15.0	2.3	5.4
2	5	6.05	0.47	2.16	33.0	2.4	10.8
3	16	3.38	0.64	1.98	54.0	10.2	32.0
4	24	7.81	0.84	5.04	187.0	20.0	121.0
5	25	3.46	1.23	2.08	86.5	30.8	52.0
6	25	4.62	1.72	2.32	115.0	43.0	58.0
7	7	4.36	1.01	2.46	30.5	7.1	17.2

* Juliano.[4]
† $K_L = K_2 H$.
Note: Ft × 0.3048 = m.

(613.5 kg/day/sq km). The absolute minimum K_L should be used for pollution permit distribution. Figure 2 shows the relationship between K_L and the allowable uniform loading.

This permissible waste loading does not depend on biochemical oxygen demand reaction rate coefficient (K_1), variations in K_2, water depths, algae growths, or organic solids deposition. None of these factors is relevant to minimum assimilative capacity. Each affects what happens in water receiving wastes but none is of any consequence in establishment of pollution policy.[6]

Summary

Much research effort has been aimed at defining the mechanism of natural reaeration and at establishing mathematical methods of predicting reaeration capacity of natural waters under various conditions. Such studies imply or state that reaeration capacity is a variable characteristic of a body of water. If this variable property is to be used as the basis of intentional pollution, then clearly the probability of the reaeration capacity's being able to meet the pollution load discharged into the body of water at any particular time should be established. This means that if the assimilative or purification capacity of a body of water is to be divided up among various polluters, regulatory agencies carrying out such a program should publish the odds on the occurrence of irreversible damage.

The maximum assimilative capacity of a stream that can be depended on to accommodate intentional pollution is equal to the minimum reaeration capacity of the stream. The practicality of the concept is obvious when one considers that all of the pollution entering a particular stream is rarely measurable. Usually only the pollution that is intentionally discharged is measured. Reserving any assimilative capacity exceeding the minimum for unmeasured pollution, such as stormwater runoff, agricultural drainage, spills, dumps, and other accidental pollution, becomes more reasonable than the present practice. One immediate benefit of acceptance of this concept is the stoppage of expensive long-term studies now under way to define assimilative capacity, and the job of cleaning up can be begun rather than delayed while these studies are carried out. The public's tax dollars are conserved, and pollution abatement can proceed.

References

1. Busch, A. W., "Use and Abuse of Natural Water Systems," *Jour. Water Poll. Control Fed.*, **43**, 1480 (1971).
2. Busch, A. W., "Biological Factors in Aerator Performance," Proc. 25th Ind. Waste Conf., Purdue Univ., Ext. Ser. **137**, 174 (1970).
3. Churchill, M. A., *et al.*, "The Prediction of Stream Reaeration Rates," *Jour. San. Eng. Div., Proc. Amer. Soc. Civil Engr.*, **88**, SA4, 1 (1962).
4. Juliano, D. W., "Reaeration Measurements in an Estuary," *Jour. San. Eng. Div., Proc. Amer. Soc. Civil Engr.*, **95**, SA6, 1165 (1969).
5. Klein, L. "River Pollution, 2: Causes and Effects," Butterworths, London, Eng. (1962).
6. Busch, A. W., "Aerobic Biological Treatment of Waste Waters—Principles and Practice," Oligodynamics Press, Houston, Tex. (1971).

IV. Prevention and monitoring

Pollution prevention

By E. L. Cronin

Pollution prevention

This paper differs from others presented in the discussion meeting in that it is a summary of information from technical literature and other sources and a consideration of practices, effects and policies. No original research is reported, and much opinion is expressed.

The magnitude of marine pollution effects is now being revealed by information from many sources. Every estuary near human concentration or activity receives the wastes of the watershed so that many are biologically degraded or destroyed. Large inshore oceanic areas, including the North Sea and the Atlantic shelf of North America, show extensive and significant evidence of the detrimental effects of human activity. In the open sea, DDT and other biocidal chemicals have been observed at Antarctica, the farthest possible point from the terrestrial sites of their use (National Academy of Sciences, etc. 1970). They reduce photosynthesis in phytoplankton, upset normal nerve functions in animals, impair vertebrate reproductive success, reduce enzymatic activity, and have additional effects. These are processes which are fundamental to the world ecosystems, and impairment is a global loss (Report Secretary's Cttee 1969).

Evidence is less available on world-wide translocation and effects of radionuclides, toxic chemicals, and other pollutants, but it is sufficient to suggest that estuaries and coastal areas are under serious threat and that oceanic systems are now or may become significantly altered.

Present magnitudes and trends

Present trends in oceanic waste disposal are not reassuring. For New York City, Gross (1970) has shown increases in the average oceanic disposal of solid wastes from 1960–3 to 1964–8 of 26 % in volume ($6.15 \times 10^6\ m^3$/year to $7.8 \times 10^6\ m^3$/year) and 26 % in mass (6.2×10^6 tonnes/year to 7.8×10^6 tonnes/year). Determination of the biological effects and total evaluation of this disposal are receiving some attention, but they have not been achieved. In the United States (Cttee on Pollution, etc. 1966) *per capita* solid waste production increased from 1.3 kg/day (2.75 lb/day) in 1920 to 2 kg/day (4.5 lb/day) in 1965. Annual *per capita* increase of about 2 % and population growth of about 2 % yield a compounding increase of 4 % and a doubling period of 20 years. Future trends are suggested by the predicted rise in the national industrial index (which will affect *per capita* production) from a level of 125 in 1954 to 767 in 2000.

Global population changes shown in table 1 demonstrate the rapid shortening of the doubling period from 1650 years to about 35 years.

Table 1. Estimated world populations at various times in recorded history

(Source: S.F.I. Bulletin, attributed to the Conservation Foundation, Washington, D.C. (from western sources).)

date	population 10^9	doubling time years
1	0.25	1650
1650	0.5	200
1850	1	50
1930	2	
1966	3.3	
		34
2000	6.6	

Various projections provide estimates of the quantity and nature of wastes for future disposal. For the Atlantic Coast of the United States, for instance, Bechtel Corporation (1969) estimates that the quantity of digested sludge from municipal sewage treatment plants will nearly triple from 1986 dry tonnes per day in 1968 to 5550 tonnes in 2000. Population growth, increased use of garbage grinding, and higher levels of waste treatment will all contribute to the increase. The 6×10^9 kg per year expected after 30 years is obviously one point on a curve which will continue to rise.

To learn the possible effects of releasing much of this material in the slope water over the continental shelf, the report includes a valuable set of estimates of the dilution requirements for six types of components of the wastes, if present biological qualities and values are to be protected (see table 2).

Each of these estimates is independent, and each contains stated assumptions and approximations. There is, however, an impressive cluster of indications that dilutions of 3×10^4 might be acceptable or essential for these factors. The total renewal rate of the continental shelf is estimated by Bechtel to be 2.26×10^{12} m^3 per year. If the predicted year 2000 quantity of 2×10^9 kg (as dry solids) of total sludge per year reaches the shelf area at a concentration of 3.5 %, the addition would be 6×10^7 m^3 per year. For 3×10^4 dilution, the required 'new' water would be 1.8×10^{12} m^3 per year. Therefore, approximately 80 % of the available capacity from new water would be required. Since wastes arise on land, the inshore waters will receive above average loading and one or more of the tolerable limits might well be exceeded. Even such a large coastal area might be useful only for a relatively short period of 20–40 years for waste disposal before destruction of other uses occurs, if present methods and materials are employed.

TABLE 2. ESTIMATED DILUTION REQUIRED TO AVOID DEGRADATION OF ATLANTIC COAST SHELF WATERS FROM DIGESTED SLUDGE AND DREDGING SPOIL

(Data from Bechtel Corp. 1969.)

contaminant	approximate concentration in sludge	estimated tolerable concentration	required dilution
coliform organisms	—	1000–3000 per ml	10^4 adequate
suspended solids	3.5 %	1 part/10^6	3×10^4 required
phosphate (PO_4)	1000 parts/10^6	0.05 part/10^6	10^4 required
nitrate (N)	1500 parts/10^6	0.5 part/10^6	2×10^3 required
BOD		no anoxia	3×10^4 adequate
heavy metals	100 parts/10^6	2 parts/10^6†	2.5×10^5 required

† In organisms, assuming biological concentration of 5×10^3.

It may, in fact, be possible to combine population projections, industrial predictions, hydrographic data and the partial information now available on biological tolerance limits to place a 'life expectancy' estimate on each body of water. We have a 40-year expectancy on the Atlantic shelf, and other systems may have a 50-year, 10-year or 0-year future for optimal or desirable biological quality. After that date, such quality and present uses will be sacrificed unless alternative waste disposal methods are brought into practice.

Four specific patterns appear to enhance the threat of pollution to the biological systems of the ocean:

(*a*) Simultaneous expansion of populations and of technological activity, producing rapid increase in total disposal quantities.

(*b*) Creation of hundreds of new chemical compounds in very large volume much faster than the growth in related understanding of their biological effects. For example, there are more than 900 biocidal chemicals, and all tests of effects combined have probably included less than 1 % of the species affected (Report Secretary's Cttee 1969).

(*c*) Tradition, and law, tend to protect the long-assumed right of the individual, community and corporation to make free and unrestricted use of the common properties of environmental air and water for waste disposal.

(*d*) Engineers, and many others, continue to assume that the ocean is a waste-receiving system of infinite capacity.

These characteristics of our present demographic, industrial, behavioural and intellectual patterns create formidable difficulties in altering the trends of marine pollution.

The significant effects of marine pollution

Deleterious introduced materials may have physical, chemical, biological or aesthetic effects. Each of these may be important in some aquatic systems, but the uses of the marine environment are significantly different from those of fresh-water lakes, streams and rivers. For instance, chemical changes in dissolved solids, pH, and other conditions directly affect potability and safety for human consumption and suitability for industrial uses in freshwater, but these uses rarely exist for marine waters. Physical changes in density, viscosity and other parameters appear to offer little threat to marine organisms. Aesthetic offence from trash, oil and odours are undesired everywhere, but do depend on the presence of the offended, who are far scarcer at sea. All of these, then, are of limited marine significance.

The biological effects of marine pollution are the most significant ones, and lie at the centre of almost all problems dealing with waste disposal. These may, obviously, be created by altering the physical or chemical system, but the importance to other persons lies in the biological result. Many types of biological impact are known to be deleterious to the ecosystem or to human uses of the ocean and several merit brief comment.

Health risks, through the dispersion and presentation of viruses, bacteria, and parasites, exist in some coastal waters (Bechtel Corporation 1969). The accumulative capacity of many shellfish for some of these organisms and the notorious concentrations of human waste in many estuaries provide extreme and ubiquitous problems of public health.

Modification of the biological system may reduce the capacity of that system for assimilation of wastes. Introduction of toxic chemicals, increase in turbidity, increment in oxygen demand or introduction of exotic species, for example, might effect such change.

Biological stimulation occurs frequently in coastal areas and will probably be achieved in the open sea (Armstrong & Storrs 1970; Nat. Acad. Sci. 1970). This results from increasing the availability of chemicals which were previously limiting factors in biological production. Considerable human advantage can be gained from the accidental or intentional enhancement of useful biological production. Conversely, far-reaching damage can occur from biostimulation which alters the ecological system by over-production of biota harmful to the ecosystem and man's uses of the sea (Nat. Acad. Sci. 1970). Eutrophication can eliminate desirable species, encourage obnoxious algae, and cause anoxic conditions from the decay of introduced materials and of the over-stimulated biota.

The marine ecosystem has evolved by the continuous selection of biological species and processes which are successful under prevailing physical, chemical, and biological circumstances (Cronin 1970). The most deleterious alterations are likely to be those which create stresses which were not within the genetic experience which has produced the species, communities and systems affected. These may kill individuals through acute toxicity or reduce species success through chronic toxicity. The community structure and efficiency may be greatly altered, especially if effects occur at critical points in the food net. The pathways and magnitude of energy flow can be reduced. Since the oceanic processes are essential to the global biogeochemical systems, it is apparent that every widespread alteration in marine biological activity is of ultimate importance.

The most significant effects of marine pollution are, then, those which deleteriously affect the marine ecosystem, man's uses of the sea, or the global roles of the oceans.

Present knowledge of effects

Recent review of the quantity and quality of data and understanding of the effects of marine pollution indicate that they are not now adequate to prevent or optimally control pollution. Every change will, of course, produce effects, and economic incentives are high for use of the sea as a receiving system. Principal gaps in essential knowledge have recently been identified for the U.S. Federal Water Quality Administration through a workshop convened by the Committee on Oceanography of the National Academy of Sciences and the Committee on Ocean Engineering of the National Academy of Engineering (Nat. Acad. Sci. Cttee on Oceanography 1970).

Participants noted that additional studies are needed for the physical processes of estuaries and coastal waters (including flux and diffusion mechanisms), the effects of decay of non-conservative components on physical factors, and the interactions between floatable and settleable materials and the physical system. Chemists recommended improved understanding of the concentrations and chemical forms of biologically significant elements and compounds; research on the chemistry of particles and processes in sediments; study of nutrient chemistry and biochemical changes; intensive research on the marine chemistry of principal wastes; and adequate chemical analysis of the marine effects of terrestrial agriculture, coastal mining and dredging and other physical changes.

Biologists expressed concern over the present inadequacy of understanding in a broad range of fundamental and practical areas. They urged intensive long-range study of the structures and dynamics of marine communities; through examination of existing waste-receiving systems to determine or suggest effects; improvement of health-risk indices; expanded research on mechanisms and significance of the biological concentration of waste which occurs in some species; improved identification of the waste tolerance limits for each use made of the sea; development and application of techniques of systems analysis and simulation through various modelling methods for marine pollution problems; and development of techniques and policies for pre-operational biological evaluation of proposed disposal programmes. None of these areas is without scientific information, but none was considered to have sufficient guidance to assure reasonable protection of the marine ecosystem and man's uses of the sea.

Is waste management achievable?

Conscious efforts are made in most parts of the world to control the volume and concentration of wastes in natural waters, including the oceans. Controls range from simple regulations to computerized management systems for large metropolitan centres. Can such management protect the optimal qualities of the marine ecosystems? Obviously, it can be of enormous value and should be vigorously attempted. At the same time, such systems are heavily dependent on a series of policies, practices and information sources which are subject to possible error. Failure might, in any one management attempt, occur if one or more of the following errors or insufficiencies is present:

1. Incorrect population prediction.
2. Erroneous estimates of the quantity or nature of industrial activity.
3. Continuation of the existing philosophy of the right to use public water for waste disposal.
4. Inadequate knowledge of the assimilation and biological effects of unknown new compounds.
5. Erroneous engineering data or calculation.
6. Insufficient understanding of the biological systems and populations affected.
7. Deficiency of funds.
8. Mechanical break-down in equipment.
9. Operational error.
10. Inadequate enforcement.
11. Weakness in legislation.
12. Political pressure.

Since each of these is known to occur, the probability that massive waste disposal will be permanently well managed appears to be very fragile indeed. At some point in time, each waste disposal problem will reach the inherent limits of the natural system. It is, therefore, most urgent that alternatives to infinite expansion of marine pollution be achieved.

Can marine pollution be prevented?

Prevention would be a most drastic change from previous and present practice. Before recent decades, each individual or group disposed of his own wastes in the 'cheapest' way possible, without regard for the public interest except for a degree of health protection. The traditional freedom to use air and water has been widely exercised. Management targets have been primarily those of removing wastes from visible sight and from politically vigorous objection (Heubeck *et al.* 1968). These objectives are still widely applied.

TABLE 3. PHENOMENA AND PROCESSES UTILIZED BY NATURE OR EXPLOITED BY ENGINEERS IN WASTE TREATMENT

(From *Waste Management and Control*, Nat. Acad. of Sci.–Nat. Res. Council 1966.)

phenomenon	status of use (1966)
1. Sedimentation	wide
2. Filtration	wide
3. Coagulation	wide
4. Chemical precipitation	wide
5. Flocculation and coalescence	wide
6. Deflocculation	none
7. Mixing, diffusion and dilution	wide
8. Flotation	wide
9. Ion exchange	special uses, expanding
10. Oxidation	wide
11. Biodegradation	wide
12. Nutrient removal	experimental
13. Molecular entrapment	none
14. Adsorption	wide
15. Desorption	very limited
16. Change of state	experimental
17. Electrostatic precipitation	wide
18. Osmosis	experimental
19. Reverse osmosis	experimental
20. Ionizing radiation	very limited
21. Evaporation	limited
22. Prevention of evaporation	experimental
23. Sterilization and disinfection	wide
24. Transpiration	none
25. Microstraining	limited
26. Synergism	none
27. Biological antagonisms	none
28. Clogging in depth	experimental
29. Magnetic processes	very limited
30. Dissolution	limited
31. Capillary action	very limited
32. Electrodialysis	wide
33. FeS precipitation	none
34. Sonics	experimental
35. Storage	wide

There are, however, many available methods for pollution control and possible prevention. The U.S. National Academy of Sciences–National Research Council (Cttee on pollution 1966) has listed 35 phenomena or processes utilized by nature or exploited by engineers in waste treatment (table 3). They have identified the usable function of each, and summarized the present stage of engineering application. About 21 were in use, and many methods appear to have potentials for new and wider application.

Approaching the estimation of the state of the art from a different point of view, the Academy–Council report lists 21 types of pollutants and indicates the status of technological capabilities for dealing effectively with each (table 4). It is noteworthy that the levels of results which are 'generally acceptable' are rapidly changing and generally rising. It is also important to recognize that *every* pollutant for which capabilities were considered to be inadequate is of biological importance. Some of those (nitrogen, trace metals, pesticides, industrial wastes and viruses) have potentials for extraordinary destruction in the marine ecosystem.

TABLE 4. CAPABILITIES OF TECHNOLOGY FOR CONTROL OF VARIOUS POLLUTANTS

(From *Waste Management Control*, Nat. Acad. Sci.–Nat. Res. Council 1966.)

pollutant	technological capability† (1966)
I. Suspended solids	
(*a*) Settleable	adequate
(*b*) Colloidal	adequate
II. Dissolved solids	
(*a*) Inorganic	
1. Total dissolved solids	available, economically limited
2. Nitrogen compounds	inadequate
3. Phosphates	available, economically limited
4. Trace metals	inadequate
5. Heavy metals	adequate
6. Acidity	adequate
7. Alkalinity	adequate
8. Radioactive elements	adequate
(*b*) Organic	
1. Biochemical oxygen demand	adequate
2. Refractory materials	
(i) Detergents	adequate
(ii) Pesticides	inadequate
(iii) Residues	inadequate
(iv) Industrial	inadequate
III. Thermal pollution	adequate
IV. Living organisms	
(*a*) Infectious agents	
1. Bacteria	adequate
2. Viruses	inadequate
(*b*) Plants	
1. Attached	available, economically limited
2. Algae	adequate
(*c*) Slimes	inadequate

† Technological capability was estimated in terms of achieving 'generally acceptable results'.

In fine, there is a wide array of tools and techniques for treating wastes, but the total present body of knowledge and application of such methods leaves the marine environment quite vulnerable to large-scale damage.

If higher levels of prevention or reduction of pollution are achieved, both costs and benefits will rise. It is, unfortunately, exceedingly difficult to place values on the benefits to be achieved from perpetuating the marine ecosystem. They are high for uses like fishing and recreational use, and perhaps incalculably high in protecting the efficiency of global biogeochemical processes. The direct costs can be more exactly assessed. The American Chemical Society has recently (1969) estimated the costs of various levels of renovation of waste water, and the equipment and operating costs are indeed high (table 5). Not all of these may be necessary for marine releases, but large-scale alternatives to the present dependence on natural processes will be expensive.

TABLE 5. ESTIMATED COSTS AND USES OF WASTE WATER RENOVATION TO VARIOUS LEVELS OF IMPROVEMENT

(From *Cleaning our environment—The chemical basis for action*, American Chemical Society 1969 (original data from Stephen & Weinberger 1968).)

level of treatment	estimated cumulative operating cost—cents/1000 gallons†	uses of treated water
primary	3.5	none
secondary (activated sludge)	8.3	non-food crops
coagulation–sedimentation	13	industry, agriculture, some recreation
carbon adsorption	17	good industrial, agriculture, ground recharge
electrodialysis	26	complete organic–inorganic control
brine disposal	33	high industrial, ground recharge
disinfection	34	potable

† Assuming 10^8 gallons/day; 1 gallon (U.S.) ≈ 3.8 litres.

NEW ALTERNATIVES

Present practices of one-directional movement and massive accumulation of wastes cannot continue indefinitely without catastrophe. The viable alternative is to recycle such materials back into re-use, preventing destructive accumulation at any point in the sequence. While technically and economically feasible systems are not now available for most marine pollution problems, several stimulating examples can be drawn from recent progress in essentially similar difficulties.

(1) In a recent symposium, Merril Eisenbud, New York City Environmental Protection Administrator, suggested recycling of newsprint to prevent pollution (*The New York Times*, 15 March 1970). The City spends about $30 per short ton* to dispose of 350 000 tons of newsprint per year. It could be salvaged and re-used for about $40 per ton. A scrap dealer can obtain only $20 per ton, but a public subsidy of $21 per ton would provide a $1 profit, save the City $9 per ton, prevent destructive disposal on land or at sea, reduce pollution from pulp mills, and lower the use of timber.

* 1 short ton (U.S.) ≈ 907 kg.

(2) The State of Maryland has recently enacted legislation to create a statewide Waste Acceptance Service to receive all municipal and industrial liquid wastes or directly supervise treatment by local governments (Heubeck *et al.* 1968). Use charges would be made, related to the volume and difficulty of treatment of wastes, and certain materials could be refused. The programme is intended to obviate the dependence of the quality of public waters upon local waste treatment programmes, which are frequently inadequate because of technical ignorance, public indifference, or limited financial resources. In addition, massive collection would permit application of scale economics which might allow reclamation and recycling of many wastes.

(3) The National Academy of Sciences–National Research Council has suggested (1966) serious study of the possibilities of deliberately combining sewage treatment operations with solid-waste disposal, water supply, and power production. The wasted heat from incineration of solid wastes or the flaring of waste gases from sewage-sludge digestion might be used in power production, with the low-quality heat-containing reject stream providing energy for treating, and perhaps distilling, the waste water available.

(4) Heat from electric generating stations is wasted in very large quantities—a 2000 MW nuclear power plant releases about 4 MW (14×10^6 Btu/h). While Scandinavians are making some use of the excess to heat houses, few other successful efforts to convert this wasted energy into a useful resource have been reported. In the United States, many efforts are being made to utilize such heat in aquaculture, and various claims are made of progress. No commercially profitable use has yet been publicly proven. The range of potentials was, however, indicated by a special subcommittee of the Maryland Thermal Research Advisory Committee (1969). About 20 suggestions included culture of shellfish and fish; biological waste treatment; agriculture; heating of homes and other buildings; ice-free roads, marinas and run-ways; warming winter recreational centres; fouling control for ships; depuration of contaminated shellfish; desalinization of salt water; and raising of deep nutrient-rich waters to the surface of the sea. In a more generalized suggestion, Mihursky (1966) has suggested that excess heat from a power plant might be used simultaneously in several parts of a nutritional cycling among people, sewage, algal production, culture of useful aquatic species and food processing.

(5) Recent news reports have described efforts by (*a*) the automobile industry to collect and re-process junked automobiles, (*b*) reusing glass as a component in 'glasphalt' highway surfacing materials, (*c*) reclamation and use of aluminium and steel beverage cans.

(6) A recent suggestion appears to contain the fascinating, although distant, potential for genuine solution of two of the world's greatest problems—the depletion of raw materials and production of mountains of waste. Eastlund & Gough (1969*a*, *b*) suggest such potentials for the 'fusion torch'. If thermonuclear reactions can be tamed in controlled fusion reactors to produce cheap electricity without

radioactive wastes, thin gases called plasmas will be produced with temperatures of hundreds of millions of kelvins. The superheated plasma would be focused on waste products which would be entirely vaporized and converted to electrified particles of their constituent elements. These could be recovered and recycled efficiently. There are many difficult technical and economic problems between present experiments and such achievement, but recent research is reported to be encouraging to the concept and its potentials.

Conclusions

Review of the rising quantities and increasing diversity and complexity of polluting wastes; of the importance of biological damage in the marine ecosystem; and of the inadequacy of scientific knowledge, technological capabilities, and governmental systems to 'manage' pollution adequately suggests a single conclusion. Pollution prevention must be achieved in the near future, at least for all large centres of population or of industrial activity, or important and even fundamental qualities of the marine system will be sacrificed over large areas.

There is some encouraging indication that re-cycling may become recognized and utilized as the only viable pattern with infinite potentials. Progress is, however, very slow indeed in view of the magnitude of the problem and its rate of increase. Remote but ultimate possibilities like those suggested for the fusion torch provide sufficient encouragement to sustain the hope that the principal values of the sea may be protected by effective prevention of destructive levels of pollution.

References (Cronin)

Armstrong, N. E. & Storrs, P. N. 1970 In *Background papers on coastal wastes management.* Nat. Acad. of Sciences Cttee on Oceanography–Nat. Acad. of Engineering Cttee on Ocean Engineering Coastal Wastes Management Study Session. In the Press.

Bechtel Corp. 1969 *Criteria for waste management,* vol. I. In *F.W.P.C.A. Waste Management Study.* Federal Water Pollution Control Adminis. of the Dept. of the Interior, May 1969.

Committee on Pollution, Nat Acad. of Sciences–Nat. Research Council. 1966 *Waste management and control.* Publ. 1400.

Cronin, L. E. 1970 In *Background papers on coastal wastes management.* Nat. Acad. of Sciences Cttee on Oceanography–Nat. Acad. of Engineering Cttee on Ocean Engineering Coastal Wastes Management Study Session. In the Press.

Eastlund, B. J. & Gough, W. C. 1969*a* The fusion torch: closing the cycle from use to reuse. Div. of Research, U.S. Atomic Energy Commission, 15 May 1969.

Eastlund, B. J. & Gough, W. C. 1969*b* The fusion torch: a new approach to pollution and energy usage. In press.

Gross, M. G. 1970 In *Background papers on coastal wastes management,* Nat. Acad. of Sciences Cttee on Oceanography–Nat. Acad. of Engineering Cttee on Ocean Engineering Coastal Wastes Management Study Session. In the Press.

Heubeck, A., Jr., Coulter, J. B., McKee, P. W., O'Donnell, J. J. & Pritchard, D. W. 1968 Program for water-pollution control in Maryland. *J. Sanitary Engineering Div., Proc. Am. Soc. Civil Engrs* 94, 283–293.

Mihursky, J. A. 1966 In *Proc. of a Seminar on Water Resources and Steam Generation of Electricity.* Water Resources Research Center, Univ. of Md and Potomac Electric and Power Co., 21 November 1966, 64–76.

Nat. Acad. of Sciences Cttee on Oceanography–Nat. Acad. of Engineering Cttee on Ocean Engineering. 1970 *Wastes management concepts for the coastal zone—Requirements for research and investigation.* Nat Acad. of Sciences–Nat. Acad. of Engineering, Washington, D.C., May 1970.

Report of the Secretary's Committee on Pesticides and Their Relationship to Environmental Health 1969 Parts I and II. U.S. Dept. of Health, Educ. and Welfare, Washington, D.C., December 1969.

S.F.I. Bulletin 1970 Population explosion. The Sport Fishing Inst., No. 212, March 1970.

Subcommittee on Environmental Improvement, Committee on Chemistry and Public Affairs. 1969 *Cleaning our environment—The chemical basis for action.* American Chemical Soc., Washington, D.C., September 1969.

The New York Times, N.Y., 15 March 1970. Eisenbud suggests recycling of newsprint to cut pollution.

Thermal Research Advisory Committee Report to the Maryland Dept. of Water Resources. 1969.

MASS–TRANSFER EFFECTS IN MAINTAINING AEROBIC CONDITIONS IN FILM–FLOW REACTORS

G. J. Kehrberger and A. W. Busch

Many experimental and theoretical models proposed to predict the response of a trickling filter to change in system parameters have been based on field data. Numerous formulations [1-3] have been developed by using curve-fitting techniques relating theoretical equations to field data by multiple regression analysis. Other theoretical models [4-9] derived from equations of continuity describe system response as a function of depth and hydraulic loading. Major parameters included in these models are organic loading, hydraulic loading, depth, and recirculation. All work has been based on the premise that an electron acceptor, oxygen, is in excess.

A different approach to modeling trickling filter systems was initiated by Atkinson *et al.*[10, 11] and supported recently by Maier.[12, 13] These workers considered a system consisting of an inclined plate reactor with the biochemical reaction occurring at a liquid-solid interface. Atkinson's experimental studies have been primarily concerned with an anaerobic environment, that is, a system in which the electron acceptor is nitrate, to eliminate the variable of interphase oxygen transport. Maier's work has been based on oxygen as electron acceptor. Swilley [14, 15] examined the two environments to a limited degree. However, a comparison of the systems was not attempted.

This discussion summarizes an experimental investigation exploring mass-transfer effects on removal of total soluble organic carbon. In the study, a rectangular inclined plate served as reactor and a soluble organic, glucose, as the primary substrate. Removal of glucose and total soluble organic carbon (TSOC) in the reactor was measured. Equations developed on a theoretical basis and tested by the experimental study have been presented elsewhere.[16]

Experimental Systems

The experimental system is illustrated in Figure 1. Flow diagrams are shown for aerobic and anaerobic experiments. The anaerobic experiments used nitrate as electron acceptor. Other workers have studied trickling filters operated without the presence of exogenous electron acceptor and have defined these systems as anaerobic.[17] For this study, "anaerobic system" implies an environment in which an exogenous electron acceptor was present but not in the form of molecular oxygen.

Anaerobic System

Feed solution in a 100-l tank (A in Figure 1) was mixed and stripped of oxygen with nitrogen gas. Stripping was continued throughout the experiment. After the first hour

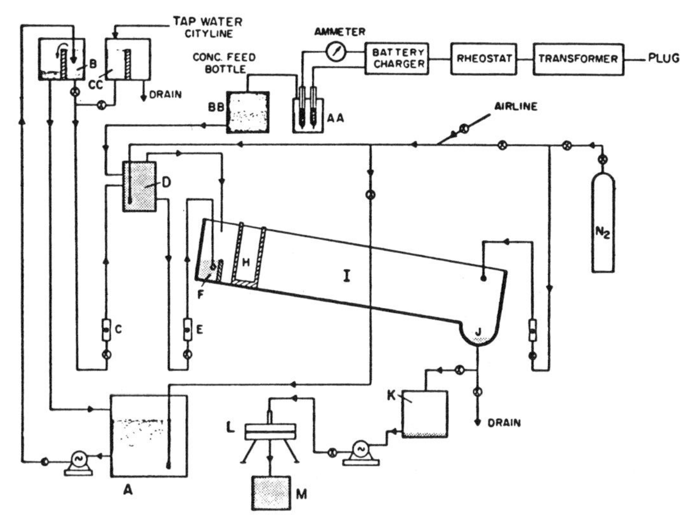

FIGURE 1.—Flow schemes for the experimental system.

of stripping, feed was pumped to a plastic tank (B) that was used as a constant-head tank and was covered with aluminum foil to minimize reaeration of feed solution. Overflow from this tank was returned to the feed storage vessel. Feed rate was controlled by a valve and a rotameter (C) as feed mixture flowed to another stripping vessel (D). This plastic bottle served as a safety precaution, eliminating any contamination by atmospheric oxygen. The vessel was maintained at a constant volume throughout the experiment. At the top of the stripping bottle, a vent line connected to the inlet area of the reactor served as a discharge line for the gases.

From the stripping vessel, feed passed through another rotameter (E) and into the inlet section of the reactor, which was also enclosed from the atmosphere by a sheet of aluminum foil. Feed leaving the weir section (F) passed through distribution section (H) and entered the reactor area (I). The reactor was covered with an aluminum plate attached securely to the sides of the reactor by long machine bolts and wing nuts. Nitrogen gas flowed continuously over the liquid surface in order to provide an inert atmosphere and to minimize intrusion of oxygen. The flowing mixture, a source of food for microorganisms, flowed over the microbial mass and entered the outlet wells (J). A small liquid reservoir was maintained at all times in the outlet wells to prevent escape of nitrogen gas. Effluent was either collected for an additional experiment or diverted into the drain system. When an additional pass of feed solution through the reactor was to be studied, the drain valve was closed and effluent was collected in a reservoir bottle (K). Liquid was then pumped continuously through a prewashed 0.45-μ membrane filter (L) into a 5-l collection bottle (M). This 124-mm diam filter removed microorganisms washed from the system.

Aerobic System

Aerobic experiments involved only a single pass through the reactor. The flow diagram, shown in Figure 1, was essentially the same as for anaerobic work except for the following obvious changes:

1. Instead of nitrogen gas, air was bubbled through the feed tank (A) and aeration vessel (D).
2. The aluminum cover plate was not used during these investigations; the reactor area (I) was left open to the atmosphere.
3. The constant-head tank (B) and inlet area (F) were not covered with aluminum foil.

Maintenance System

When experiments were not being performed, a continuous feed technique was used to maintain the growing microbial mass on the reactor. A concentrated feed solution summarized in Table I was prepared each day.

An electrolytic cell (AA in Figure 1) containing 2 percent sulfuric acid regulated the flow of concentrate from the feed bottle (BB) to the stripping vessel (D). Tap water was supplied to the system from a cold-water line passing initially through a constant-head tank (CC). During maintenance periods the stripping vessel (D) also served as a mixing vessel, blending the concentrated feed and tap water to the appropriate concentration.

After initial studies in which an aerobic environment was investigated, anaerobic experiments were conducted in more detail. Additional aerobic studies concluded the project. After changes in environment, acclimation of the microorganisms was carried out for at least 1 month. When experiments were not conducted, environmental conditions identical to experimental conditions were provided by using the maintenance system.

TABLE I.—Nutrient Mixture

Component	Proportion	
	Anaerobic System	Aerobic System
Glucose	base	base
$MgSO_4 \cdot 7H_2O$	0.5 g/g glucose	0.5 g/g glucose
$NaH_2PO_4 \cdot H_2O$	0.5 g/g glucose	0.5 g/g glucose
KNO_3	3.38 g/g glucose	0
NH_4Cl	0.21 g/g glucose	0.21 g/g glucose
$NaMoO_4 \cdot 2H_2O$	0.74 mg/l	0.75 mg/l

Reactor

The principal component of the experimental system shown in Figure 2 was the reactor, comprising the inlet, outlet, and reaction regions. A detailed description of the entire reactor apparatus has been reported elsewhere.[16] The most important section was the reaction zone. The reactive area, constructed from an acrylic plastic plate 9 in. (23 cm) wide and 38 in. (97 cm) long, was sanded uniformly in the lateral direction. The roughened surface was provided so that microorganisms could easily be grown and maintained. Sample ports, 0.5 in. (1.3 cm) in diam, were constructed along the length of the plate, positioned 3, 6, 12, 18, 24, 30, and 36 in. (7.6, 15, 30, 46, 61, 76, and 91 cm) from the inlet. Throughout the experimental study, the plate was inclined at 5 deg from the horizontal.

Sampling

Samples were gathered at the sample ports in the following manner. A port plug was removed, and the sample probe shown in Figure 3 was inserted into the hole. The stainless steel probe was fitted flush with the surface of the plastic plate and feed solution was diverted from the main stream by the knife-edged semitubular head of the sampler. A test tube was fastened to the device, and a 25-ml sample was collected. Samples were filtered through a prewashed 0.45-μ membrane filter before analysis for soluble organic carbon and glucose.

Experimental Procedures

Experiments were carried out to investigate removal of total soluble organic carbon and glucose for two mass-transfer situations (that is, aerobic and anaerobic environment). In addition,

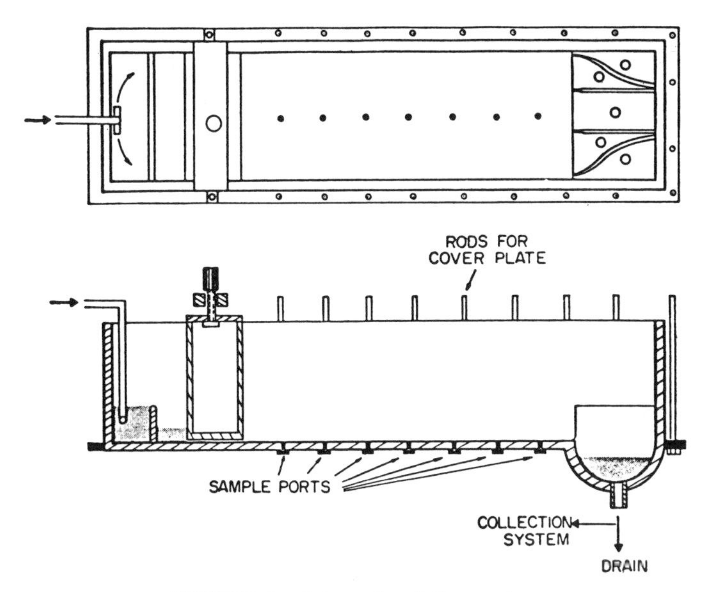

FIGURE 2.—Schematic of the reactor.

performance of the two systems was evaluated for different values of Reynolds number and substrate concentration. For the first pass of feed mixture through the reactor, samples were collected at the nine sampling stations. In addition, effluent from the anaerobic system was collected and again passed through the reactor after membrane filtration. Samples were gathered at every other port. These aliquots were analyzed for glucose and TSOC. Glucose was determined with an enzymatic colorimetric technique.[16] TSOC was monitored using a Beckman carbonaceous analyzer.

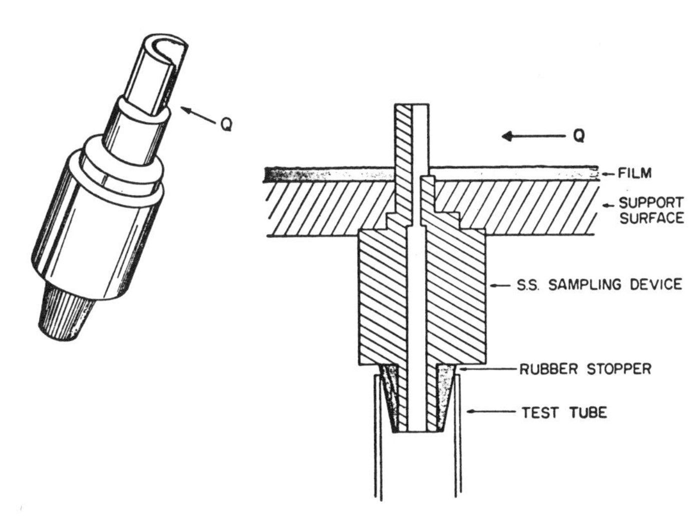

FIGURE 3.—Schematic of sample probe.

Anaerobic Experiment

Preparation for an Experiment:—Approximately 2 days before each experiment, a periodic check of volumetric flow rates and the homogeneity of the microbial film was initiated and continued throughout the experiment to insure that uniform conditions would be maintained during the study. The night before an experiment, tubing, stripping vessel, rotameters, and other pertinent equipment were replaced with clean parts. The inlet weir area was cleaned thoroughly and flushed with tap water to reduce the possibility of any microbial growth during the course of the experiment. The aluminum cover plate was removed, and the reactive area was brushed gently with the scraper to remove any unevenness on the film. The reactive surface was carefully checked for any obvious signs of non-uniformity, and the outlet section was washed thoroughly with cleaning solution. Inlet and exit tubing was replaced with clean tubing. The cover plate was restored, and the existing feed solution was replaced with a fresh mixture. The system was operated overnight under conditions identical to those of the proposed experiment. A constant-head tank, stripping vessel, clean tubing, and other fittings that were to be used as part of the experimental system were placed parallel to the maintenance system so that a quick change could readily be made in the experimental setup.

Experimental Procedure:—A 100-l feed tank was filled with a mixture of tap water and necessary nutrients. The concentration of nutritive components is presented in Table I. Concentration of the carbon source, glucose, was approximately 200 mg/l for the first pass of each experiment. While the feed solution was mixed and stripped, the reactor was checked for any irregularities. The cover plate was removed and the film was checked for any non-uniform microbial growths. The inlet area was investigated also for any biological activity and, after smoothing the film and cleaning the outlet area, the reactor plate was checked for any distorted hydraulic conditions. The aluminum plate was then replaced and sealed to the reactor section. Checking the reactor required approximately 15 min. After 1 hr of stripping, the centrifugal pump was started and feed was passed through the constant-head tank and returned to the feed tank. At this time, the maintenance system was turned off by closing the discharge valve from the constant-head tank and the experimental system was put into operation. The liquid that was stored behind the weir was siphoned off quickly and was replaced immediately with new feed mixture. Flow rates were checked to insure proper operating conditions, and the system was watched for any disturbances for at least 1 hr while the reactor returned to steady state. After steady state was reached, a sample was taken at each port starting at the outlet and progressing up the plate. Then hydraulic conditions were checked and the sampling technique was repeated.

At least two samples were collected at each port; however, for experiments with Reynolds numbers having values less than 50, three samples were collected at each port. Also at this time the three central exit streams diverted from the drain lines were filtered and collected in 5-l bottles. Completion time of the first-pass experiments varied from 5 to 10 hr, depending on the Reynolds number of the system. The standby maintenance system was put into operation at the end of the first phase of the study. Bottled effluent having an accumulated volume of approximately 40 l was mixed in a 100-l feed tank and stripped in the same manner as in the initial phase of the study. The remaining 60 l consisted of flow along the sides of the reactor and approximately 15 l of unused feed solution.

The same procedure was followed for the second pass through the reactor, with the exception that a sample taken at every other port was repeated only once because of the limited volume of feed solution. The total time of the two-pass experiments ranged from 7 to 15 hr.

The aerobic procedure was identical to that of the anaerobic studies except that samples for glucose determination were not passed through an ion exchange column. Also, only one-pass experiments were studied because the objective of the investigation was to observe the relative rates of removal between soluble organic carbon and glucose. An obvious physical difference in the two systems was the putrefactive odors that were noticed during all aerobic experiments. The aerobic system was easier to operate and control because there was no need to isolate the system from the atmosphere.

Clearly, in order to maintain an anaerobic environment, dissolved oxygen (DO) must not be present. Precautionary measures to prevent oxygen intrusion have been discussed previously. Periodic tests for DO were made using the Alsterberg modification of the Winkler Method.[18]

Throughout the experimental study, a steady state with respect to the production and removal of organic material was maintained by establishing uniform flow rates and a constant reactive surface area. Frequent smoothing of the biological film helped to achieve these uniform conditions. The system was assumed to have reached steady state when repeated measurements of glucose at a given point in the reactor did not vary with time. Preliminary investigations indicated that steady state was attained less than 1 hr after the system was put into operation, as shown in Figure 4.

Study was made of a flow regime of fully developed smooth laminar flow. Reynolds numbers at which turbulence is observed in liquid films have been tabulated by Fulford,[19] and the smallest reported value is 144. The maximum Reynolds number investigated in this study was 120, with values in most experiments ranging between 20 and 50. Wavy laminar flow or rippling of the liquid film has been reported by Fulford [19] to begin at a Froude number equal to one. Experiments reported

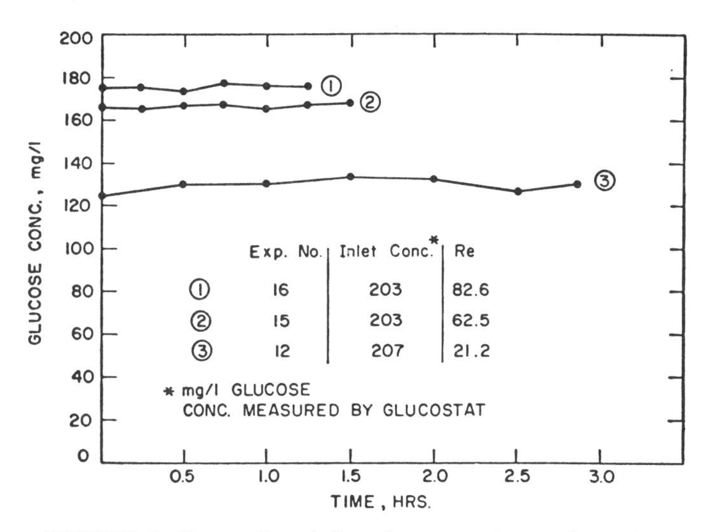

FIGURE 4.—Temporal variation of concentration at the outlet.

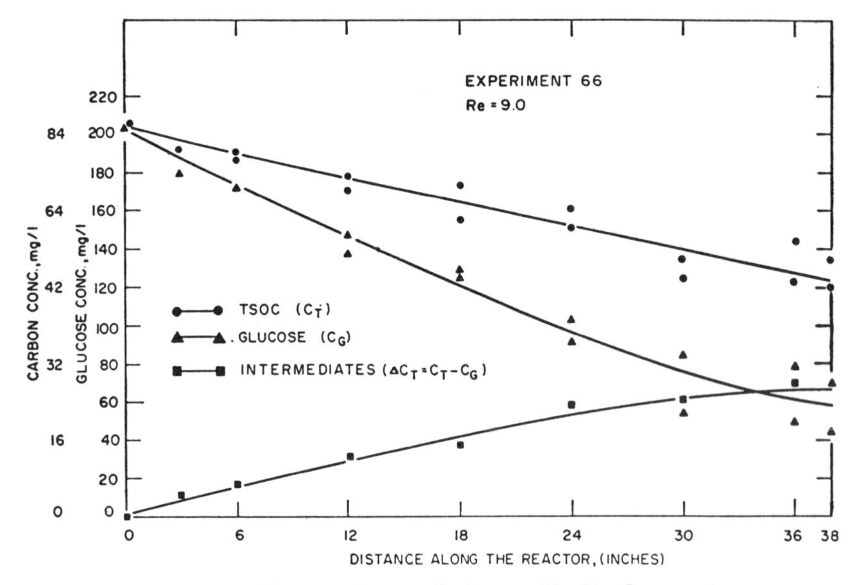

FIGURE 5.—Concentration profile for aerobic film-flow reactor.

here were carried out with Froude numbers less than one. Reflected light images from the laminar liquid film supported the premise that rippling was at a minimum.

Results

Continuous-flow experiments were made to study the removal of TSOC and glucose under aerobic and anaerobic conditions. The basic difference

TABLE II.—Summary of Aerobic Experiments*

Experiment No.	Date	N_R	Temperature (°C)	C_{Gi} (mg/l)	C_{Go}* (mg/l)	f_G	C_{Ti} (mg/l)	C_{To} (mg/l)	f_T
3	10/11/66	119.0	25.0	251	225	0.895	251	242	0.965
4	10/13/66	123.0	26.5	193	186	0.963	193	184	0.953
5	10/14/66	111.0	26.3	209	197	0.945	209	202	0.970
6	10/15/66	42.0	25.8	185	184	0.995	185	174	0.940
7	10/18/66	16.0	25.0	205	145	0.675	—	—	—
8	10/20/66	25.3	25.3	199	158	0.795	199	165	0.830
9	10/22/66	23.8	25.5	202	144	0.713	202	148	0.738
10	10/26/66	53.0	25.8	162	136	0.840	162	145	0.895
11	10/31/66	23.8	25.2	203	101	0.498	203	131	0.645
12	11/02/66	21.2	24.3	207	129	0.623	207	164	0.793
13	11/04/66	26.4	24.7	198	146	0.738	198	167	0.845
14	11/07/66	63.5	25.3	192	151	0.785	192	176	0.915
15	11/09/66	62.5	25.3	203	166	0.818	203	185	0.910
16	11/11/66	82.8	24.9	203	175	0.863	203	194	0.955
66	04/21/68	9.0	24.8	203	54	0.266	205	120	0.585
67	04/21/68	9.0	25.3	205	67	0.326	205	128	0.625
68	04/27/68	17.9	25.5	210	147	0.700	210	204	0.970
69	04/28/68	29.3	25.5	201	147	0.732	198	166	0.839
70	04/28/68	22.4	26.5	201	142	0.708	201	176	0.876

* Symbols are defined in the section entitled "Nomenclature" preceding "References."
† Average values, concentrations in terms of glucose.

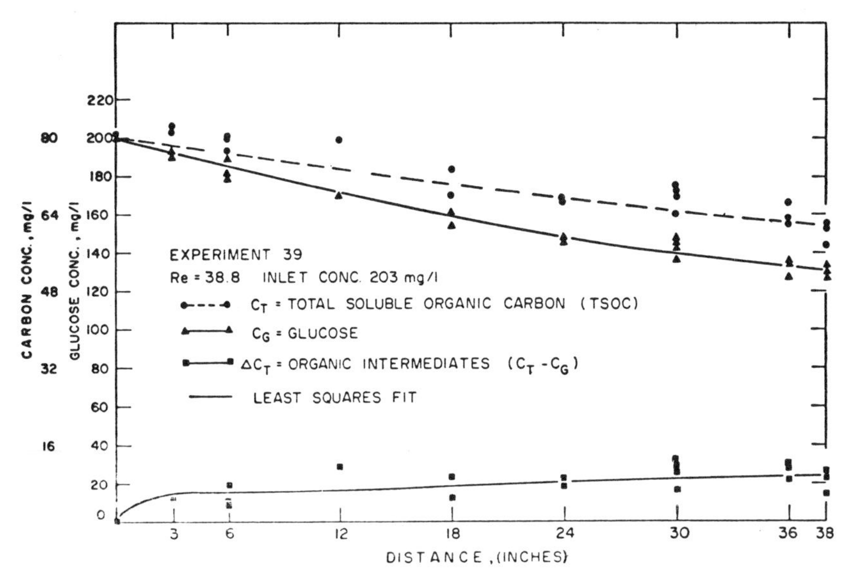

FIGURE 6.—Concentration profile for anaerobic film-flow reactor.

between the two experimental systems was that electron acceptor was present in the liquid phase for the nitrate (anaerobic) system, while interphase transport of oxygen was critical in the aerobic reaction. A detailed consideration of reactor response to variables such as Reynolds number and substrate concentration was made previously.[20]

A typical output observed for an aerobic experiment is shown in Figure 5.* Glucose and TSOC were monitored along the length of the reactor. A buildup of organic intermediates was observed in these experiments. The aerobic study explored influent carbon concentrations ranging from 64 to 100 mg/l and Reynolds numbers varying from 9 to 120. Measured concentrations of intermediates were as low as 1 mg/l and as high as 26 mg/l organic carbon. Table II summarizes the results from the aerobic experiments.

Profiles observed for anaerobic experiments were similar to those from aerobic tests. A typical response is depicted in Figure 6. A decrease in glucose and TSOC and an increase in intermediates was observed for influent organic carbon concentrations ranging from 52 to 86 mg/l. Reynolds numbers varied from 20 to 100. Intermediates concentration ranged from 1 to 18 mg/l organic carbon.

Effluent from the anaerobic system was membrane-filtered, collected, and again passed through the reactor. A typical output is shown in Figure 7. Concentration of glucose decreases rapidly from an initial (first-pass) value of 88 mg/l (as organic carbon) to a value of 14 mg/l (as organic carbon) at the end of the second pass. TSOC, removed at a lower rate than glucose, reached a final concentration of 32 mg/l from the initial 88 mg/l. Profiles show an initially rapid decrease in TSOC and glucose during the first pass. Curves then leveled off, indicating the effect of a vertical con-

* The data given in Figures 5, 6, and 7 are reported in terms of organic carbon and glucose. The two scales are presented so that a common basis could be used to evaluate the amount of carbon associated with the organic intermediates. Therefore, the glucose values can be converted to organic carbon values by multiplying the glucose concentration by 0.4.

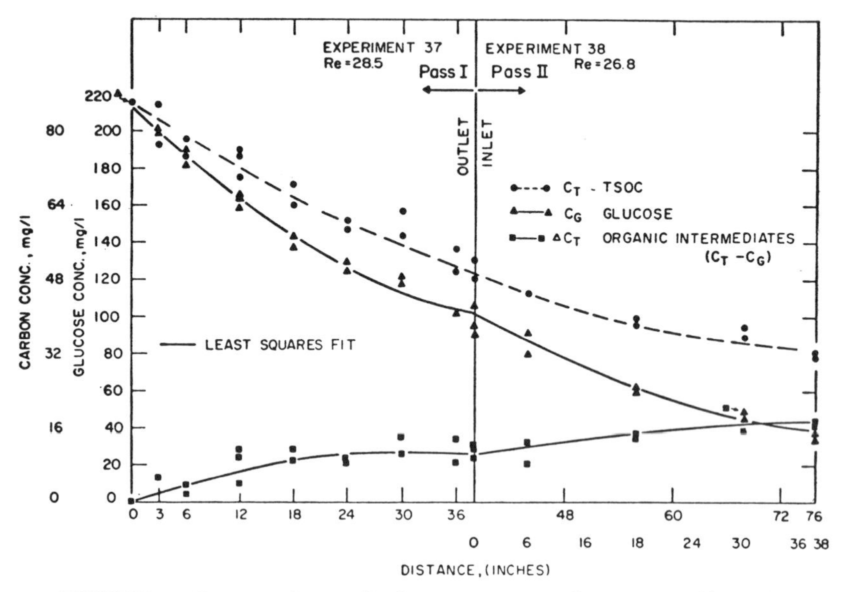

FIGURE 7.—Concentration profile for two-pass experiment, anaerobic study. (In. × 2.54 = cm.)

centration gradient on removal of organic material. Mixing the exit stream destroyed the gradient, and passing the effluent through the reactor for a second time resulted in an initial increased removal rate of both TSOC and glucose. A continuous buildup of organic intermediates was always noted. In some cases the final concentration of intermediates was greater than the effluent glucose carbon concentration. The results from the anaerobic experiments are presented in Table III.

Similar responses were observed during the aerobic study, as depicted by the output shown in Figure 5. Glucose and TSOC profiles leveled off in the last third of the reactor and, in some cases, concentration of carbon in intermediates was greater than glucose remaining.

Mass-transfer effects on removal of organic substrates for this experimental system are depicted in Figures 8 and 9, which show the fraction remaining of glucose and TSOC as a function of Reynolds number. For the same Reynolds number, a higher fraction of removal of both TSOC and glucose was accomplished in the anaerobic (nitrate) system. In addition, the curves demonstrate that an increasing Reynolds number caused a decrease in removal of organic substrates.

Discussion

This study examined the removal of soluble organic compounds through a biological reactor with the biochemical reaction taking place at the liquid-solid interface. Results demonstrated that in addition to glucose, other soluble organic compounds were present in the liquid phase for both aerobic and anaerobic (nitrate) systems.

The basis for using different electron acceptors in a study of mass-transfer effects is work done by Schroeder [21] in a completely mixed system. Schroeder measured the production of organic intermediates in a nitrate system and observed that when the primary substrate, glucose, had disappeared his mixed bacterial popula-

TABLE III.—Summary of Anaerobic Experiments

Experiment Number	Date	N_R	Temperature (°C)	$\delta \times 10^2$ (cm)	X	C_{Gi}* (mg/l)	C_{Go}* (mg/l)	f_{Go}†	C_{Ti}* (mg/l)	C_{To}* (mg/l)	f_T†	C_{Ii}* (mg/l)	C_{Io}* (mg/l)	f_I†
	(1967)													
35	12/08	38.8	25.5	2.98	0.201	209	126	0.604	209	129	0.618	0.00	3.0	0.014
36	12/08	32.8	24.0	2.82	0.251	122	64	0.525	124	80	0.645	2.00	16.0	0.131
37	12/12	28.5	26.0	2.68	0.305	215	93	0.434	215	123	0.576	0.00	30.0	0.140
38	12/12	26.8	25.5	2.63	0.328	106	37	0.349	130	80	0.615	24.00	43.0	0.405
39	12/16	38.8	25.0	2.98	0.201	203	130	0.640	203	151	0.745	0.00	21.0	0.103
40	12/16	28.3	23.5	2.68	0.305	119	59	0.495	144	81	0.632	25.00	32.0	0.269
41	12/20	43.3	25.0	3.09	0.174	204	129	0.632	204	163	0.800	0.00	34.0	0.167
42	12/20	41.8	23.0	3.10	0.182	130	72	0.555	160	116	0.725	30.00	44.0	0.338
43	12/24	50.5	25.0	3.26	0.140	202	139	0.689	202	162	0.802	0.00	23.0	0.113
44	12/24	40.1	24.0	3.00	0.191	139	85	0.612	157	110	0.700	18.00	25.0	0.180
	(1968)													
45	01/13	32.8	24.0	2.82	0.251	196	112	0.571	196	133	0.678	0.00	21.0	0.107
46	01/13	23.8	23.0	2.53	0.384	116	43	0.371	137	70	0.511	21.00	27.0	0.233
47	01/17	52.2	25.0	3.29	0.135	212	136	0.640	212	163	0.768	0.00	27.0	0.127
48	01/17	47.7	24.0	3.19	0.152	138	85	0.615	159	117	0.735	21.00	32.0	0.232
49	01/22	74.5	26.0	3.70	0.084	209	163	0.780	209	182	0.870	0.00	19.0	0.091
50	01/22	59.6	24.5	3.44	0.113	156	113	0.725	178	149	0.838	22.00	36.0	0.231
51	01/24	101.0	26.0	4.10	0.056	209	178	0.852	209	185	0.885	0.00	7.0	0.034
52	01/24	98.5	24.5	4.06	0.058	170	144	0.848	185	176	0.950	15.00	32.0	0.180
53	01/29	97.0	26.5	4.00	0.059	213	174	0.816	213	177	0.831	0.00	3.0	0.014
54	01/29	86.5	25.0	3.89	0.069	172	138	0.803	177	166	0.938	5.00	28.0	0.163
55	01/31	83.5	26.5	3.85	0.072	207	166	0.802	207	154	0.745	0.00	−12.0	—
56	01/31	62.5	25.0	3.49	0.106	157	119	0.758	151	126	0.835	−6.00	7.0	0.045
57	02/07	46.2	24.8	3.16	0.159	206	140	0.680	206	152	0.738	0.00	12.0	0.058
60	02/26	23.8	25.0	2.53	0.384	211	114	0.540	211	129	0.612	0.00	15.0	0.071
63	03/06	23.8	25.0	2.53	0.384	204	99	0.485	204	126	0.525	0.00	27.0	0.132

* Average value; concentrations are in terms of glucose.

† $f_T = C_{To}/C_{Ti}$; $f_G = C_{Go}/C_{Gi}$; $f_I = C_{Io}/C_{Gi}$.

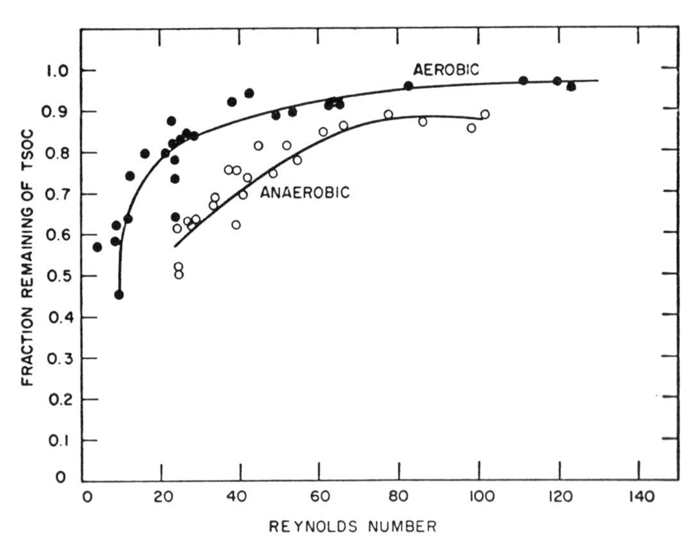

FIGURE 8.—Comparison of carbon removal (TSOC) for anaerobic and aerobic experiments.

tion used the excreted intermediates as their source of carbon.

However, Schroeder[21] did not observe the presence of organic intermediates in his mixed aerobic systems. His batch reactors contained an excess of exogenous electron acceptor, oxygen, so that organisms could take glucose to terminal oxidation by appropriate metabolic pathways. In addition, the organisms were a mixed bacterial population in which certain species could utilize the waste products released by other organisms. For the aerobic film-flow reactor, the presence of organic intermediates could be the result of

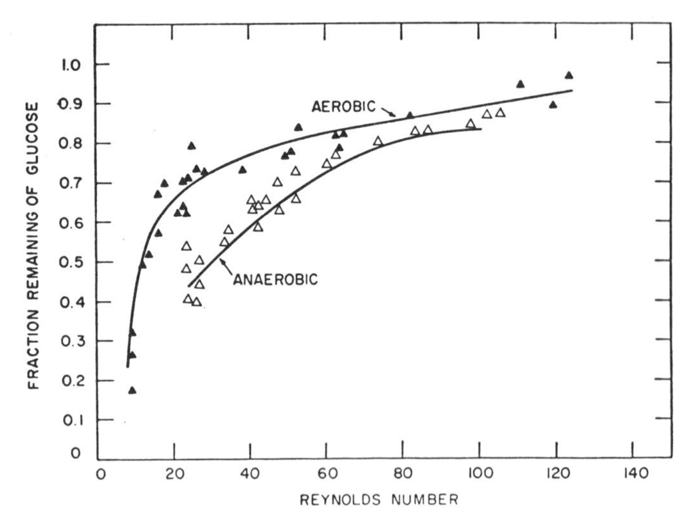

FIGURE 9.—Comparison of glucose removal for aerobic and anaerobic experiments.

either an oxygen-limited environment or a selective microbial population.

Schroeder[21] demonstrated that for a mixed batch reactor using either an exogenous or endogenous electron acceptor and oxygen- or nitrate-acclimated seed, the rate of glucose removal was essentially the same. His results supported the concept of glycolysis being the principal metabolic pathway in the oxidation of glucose. For the heterogenous film-flow reactor, the biomass may use entirely different biochemical pathways for the anaerobic and aerobic environments. For example, the pentose shunt is another pathway used to initiate the terminal oxidation of glucose. Therefore, if the biochemical pathways were different for the two systems, the glucose removal rates probably would not be equal, as indicated by the experimental data.

A bacterial population acclimated to a specific environment for a long time may become selective in that only certain organisms will survive the acclimation. Such a biomass may lack the ability to take glucose completely through the Krebs cycle, causing the buildup of organic intermediates.

Comparison of aerobic and anaerobic studies depicted in Figures 8 and 9 clearly demonstrates that removals of both glucose and TSOC are greater in the anaerobic (nitrate) system. The microorganisms in the nitrate system had an excess of exogenous hydrogen acceptor and metabolized glucose to a greater degree than those in the aerobic system. This resulted in the removal of a larger amount of substrate. The higher level of TSOC observed in the aerobic system is probably the result of incomplete glucose metabolism, which caused a large amount of intermediates to be produced.

Because putrefactive odors characteristic of septic systems were released during the nominally aerobic film-flow studies, the microbial population may have been oxygen-limited and therefore restricted in carrying out normal aerobic oxidation of glucose. An oxygen-limited environment would cause an interruption in the Krebs cycle, resulting in a buildup of organic intermediates.

This condition was the result of the type of hydrodynamic system studied. Although the feed to the aerobic system contained DO, metabolism of organic substrates caused a rapid oxygen depletion and the subsequent release of objectionable odors. In the smooth, fully developed, laminar flow regime studied, the lack of rippling at the gas-liquid interface controlled the transport of oxygen across the interface. Wave motion enhances oxygen transfer from the gas to the liquid phase by increasing surface area and providing a mixing zone at the surface. A minimum Reynolds number of 144 is required to insure a rippling condition. However, for higher Reynolds numbers, a substantial removal of TSOC could not be observed in this experimental system because of the short reaction zone. Experiments carried out at Reynolds numbers greater than 100 accomplished only a small change in organic content from the inlet to the outlet.

In addition, rippling does not guarantee complete mixing throughout the liquid phase. Turbulent flow regimes could be investigated with a longer plate to study the effect of concentration gradients on organic removal. However, short contact time and sloughing of microorganisms are common problems associated with high velocities of flow.

Oxygen transport can be increased in field installations that employ vertical synthetic packing material by recirculating the clarified effluent. Turbulent conditions would cause an increase in both oxygen transfer and the mass transfer of organic substrate. Recirculating an effluent containing microorganisms would produce a higher concentration of organisms and would accomplish a greater removal of organic materials. These considerations indicate, however, that a simpler approach in designing waste treatment facilities is to select completely mixed reactors instead of static trickling filters.

Another method for increasing mass transport is to design the system so that the liquid flowing over the media is internally mixed. Stacking the media so that volumetric flow is not uniform throughout the reactor and providing corrugated packing material that causes a "rolling" condition produces localized mixing that increases mass transfer of oxygen. In addition, concentration gradients are destroyed and the system is not controlled by this mechanism. Liquid-phase transport limitations were commonly observed in the experiments carried out in this study.

Conclusions

1. An experimental study of a laminar film-flow biological reactor showed that mass-transfer effects can result in decreased effectiveness of the biological reaction.

2. Under nominally aerobic conditions mass-transfer effects were manifested by production of putrefactive odors and buildup of organic intermediates in the liquid phase.

3. An intentionally anaerobic system in which nitrate was used as electron acceptor eliminated the need for interphase transport of oxygen and accomplished a greater removal of organic carbon, both primary substrate and the characteristic intermediates.

4. Interphase transport of reactants to the reaction site may have produced the difference in reaction rates in the two systems, but biochemical factors may also be important.

5. Full-scale trickling filter installations should provide turbulent conditions to minimize transport effects and should recirculate unsettled effluent to offset the decreased contact time resulting from high flow velocities. The logical extension of this argument is a continuous-flow stirred tank reactor.

Acknowledgment

The work reported in this paper was supported in part by Graduate Training Grants WP-12 and WP-44 from the FWPCA.

Nomenclature

C_{Gi} = influent glucose concentration
C_{Go} = effluent glucose concentration
C_{Ii} = influent intermediates concentration
C_{Io} = effluent intermediates concentration
C_{Ti} = influent TSOC concentration
C_{To} = effluent TSOC concentration
f_G = fraction glucose remaining
f_I = fraction intermediates remaining
f_T = fraction TSOC remaining
N_R = Reynolds number
R = temperature, °C
X = dimensionless length
δ = liquid film thickness

References

1. Eckenfelder, W. W., "Industrial Water Pollution Control." McGraw-Hill Book Co., Inc., New York, N. Y. (1966).
2. Galler, W. S., and Gotaas, H. B., "Analysis of Biological Filter." *Jour. San. Eng. Div., Proc. Amer. Soc. Civil Engr.*, **90**, SA6 (1964).
3. Velz, C. J., "Basic Law for Biological Filters." *Sew. & Ind. Wastes*, **20**, 615 (1948).
4. Ames, W. F., *et al.*, "Transient Operation of the Trickling Filter." *Jour. San. Eng. Div., Proc. Amer. Soc. Civil Engr.*, **88**, SA3 (1962).
5. Germain, J. E., "Economical Treatment of Domestic Waste by Plastic-Medium Trickling Filters." *Jour. Water Poll. Control Fed.*, **38**, 192 (1966).
6. Howland, W. E., "Flow over Porous Media as in a Trickling Filter." *Proc. 12th Ind. Waste Conf.*, Purdue Univ., Ext. Ser. **94** (1957).
7. Kornegay, B. H., and Andrews, J. F., "Kinetics of Fixed Film Biological Reactor." *Proc. 22nd Ind. Waste Conf.*, Purdue Univ., Ext. Ser. **129** (1967).

8. Schulze, K. L., "Load and Efficiency of Trickling Filters." *Jour. Water Poll. Control Fed.*, **32**, 245 (1960).
9. Ames, W. F., and Collings, W. Z., "Diffusion and Reaction in the Biota of the Trickling Filter." Tech. Rept. No. 19, Dept. Mech. Eng., Univ. of Delaware, Newark (Oct. 1962).
10. Atkinson, B., *et al.*, "Recirculation, Reaction Kinetics, and Effluent Quality in a Trickling Filter Flow Model." *Jour. Water Poll. Control Fed.*, **35**, 1307 (1963).
11. Atkinson, B., *et al.*, "Kinetics, Mass Transfer, and Organism Growth in a Biological Film Reactor." *Trans. Inst. Chem. Engr.*, **45**, T257 (1967).
12. Maier, W. J., *et al.*, "Simulation of the Trickling Filter Process." *Jour. San. Eng. Div., Proc. Amer. Soc. Civil Engr.* **93**, SA4, 91 (1967).
13. Maier, W. J., "Model Study of Colloidal Removal." *Jour. Water Poll. Control Fed.*, **40**, 478 (1968).
14. Swilley, E. L., Ph.D. thesis, Rice Univ., Houston, Tex. (1964).
15. Swilley, E. L., and Atkinson, B., "A Mathematical Model for the Trickling Filter." *Proc. 18th Ind. Waste Conf.*, Purdue Univ., Ext. Ser. **115** (1963).
16. Kehrberger, G. J., and Busch, A. W., "The Effect of Recirculation on the Performance of Trickling Filter Models." *Proc. 24th Ind. Waste Conf.*, Purdue Univ. (1969).
17. McCarty, P. L., "Anaerobic Treatment of Soluble Wastes." In "Advances in Water Quality Improvement." E. F. Gloyna and W. W. Eckenfelder, Jr. [Eds.], Univ. of Texas Press, Austin (1968).
18. "Standard Methods for the Examination of Water and Wastewater." 11th Ed., Amer. Pub. Health Assn., New York, N. Y. (1960).
19. Fulford, G. D., "The Flow of Fluid in Thin Films." In "Advances in Chemical Engineering." Vol. 5, Academic Press, New York, N. Y. (1964).
20. Kehrberger, G. J., Ph.D. thesis, Rice Univ., Houston, Tex. (1968).
21. Schroeder, E. D., "Dissimilatory Nitrate Reduction by Mixed Bacterial Populations." Ph.D. thesis, Rice Univ., Houston, Tex. (1966).

USE OF POLYELECTROLYTES IN TREATMENT OF COMBINED MEAT–PACKING AND DOMESTIC WASTES

K. D. Larson, R. E. Crowe, D. A. Maulwurf, and J. L. Witherow

Organic and inorganic flocculants have been used for a number of years in wastewater treatment practices for increased sedimentation of solids and for increased efficiency in sludge filtration. These practices have mainly been in the area of domestic waste treatment. A waste treatment problem involving 90 percent meat-packing waste has evolved over the years at the South St. Paul municipal plant. This problem precipitated an investigation of the use of chemicals to improve the treatment efficiency and decrease the related cost. The use of chemicals, especially a dual system of organic and inorganic chemicals, to treat this combined waste was considered to be a unique feature of the investigation.

History

In 1940, South St. Paul, Minn., completed a 10-mgd (37,850-cu m/day) high-rate, trickling filter plant to treat combined domestic and meat-packing wastes. The strength of the waste increased from 445,000 population equivalents/day in 1940 to 917,000 population equivalents/day in 1964. These figures for killing days were reduced to approximately 10 percent on weekends. During this period of years the efficiency of the plant dropped from 72 percent to 58 percent removal of biochemical oxygen demand (BOD).

In 1958 it became apparent that additional treatment would be required by both the adjacent Minneapolis–St. Paul sanitary district plant and the South St. Paul plant. The primary concern at that time was the low levels of dissolved oxygen in the Mississippi River during the late summer months.

Pilot-plant studies conducted at South St. Paul indicated that the plant effluent BOD could be reduced by more than 50 percent by passing the effluent through an anaerobic pond of sufficient capacity to afford at least a 5-day detention period. This addition was initiated in February 1962.

During the 3-yr period, April 1, 1959–March 31, 1962, the average BOD in the South St. Paul effluent was 464 mg/l during killing days. The state of Minnesota required 65 mg/l BOD effluent quality, which was anticipated as matching that of the Minneapolis–St. Paul sanitary district effluent when its new (then under bid) high-rate activated sludge plant would be completed. The anaerobic pond reduced the BOD of the normal plant effluent by 66 percent, but this still left an effluent of approximately 150 mg/l, 2.5 times stronger than the 65 mg/l required. Consequently, the city of South St. Paul engaged an engineering firm to study the loads and capacity of the plant and to recommend a program to meet the BOD requirements of 65 mg/l and a method of disposal of sludge other than lagooning.

Shortly after the engineering report was presented to the city, the Upper Mississippi River Pollution Abatement Conference was held jointly by Wisconsin, Minnesota, and the Federal Water Pollution Control Administration (FWPCA). On completion of the conference the limit of 65 mg/l BOD in the plant effluent was lowered to 50 mg/l and subsequently by the state of Minnesota to 35 mg/l.

In February 1965, the engineering firm developed several new plans encompassing second-stage filtration and incineration for a total cost of approximately $10 mil. The packing industries contracted an engineering firm to review this report, which recommended treating the hot, packing plant wastes by the anaerobic contact process. The cost again approached $10 mil for the entire package, an unacceptable figure from the viewpoint of the industries.

The industries became interested in the possible efficiency and economy of chemical treatment of the meat-packing wastes. A review of the literature and existing patents indicated that the use of chemicals might reduce construction cost by $2 mil. The city of South St. Paul made application to FWPCA for a demonstration grant entitled, "Efficiency and Economy of Polymeric Sewage Clarification," to determine the effectiveness of a chemical flocculation process on combined domestic and packing plant wastes or on packing plant wastes alone. New grit chambers were necessary along with a new intercepting sewer from the industries and a new pumping station to separate the meat-packing wastes. The increased costs of new grit chambers, chemical feed building, flash mixers, feed equipment, chemicals, and personnel to undertake the demonstration were estimated at $845,000. The FWPCA agreed to participate at 55 percent of the cost, not to exceed $450,000. The city of South St. Paul and the packing industries agreed to fund the remainder of the project cost.

Plant Description

Even without the anaerobic stabilization pond, the South St. Paul wastewater treatment plant is not a conventional trickling filter plant (Figure 1).

The wastewater enters the plant, passes through a 1-in. (2.54-cm) bar screen, and is pumped into three grit tanks. At design flow of 10 mgd (37,850 cu m/day), the detention time in these tanks is 30 min. During the killing hours at the meat-packing plants, the flow rate will be 20 mgd (75,500 cu m/day), thereby reducing the detention time by 50 percent. From the grit tanks the wastewater flows through mechanical flocculators. The detention times at 10 mgd (37,850 cu m/day) in the flocculation tanks and primary clarifiers are 45 min and 1.5 hr, respectively. Between the mechanical flocculators and the standard rectangular primary clarifiers is a baffle wall. The primary effluent goes to six trickling filters and then to the final clarifiers, which have a detention time of 1.5 hr at 10 mgd (37,850 cu m/day). The secondary effluent is pumped to the anaerobic ponds with 5 days of detention and then discharged to the Mississippi River. The sludges from the primary and secondary clarifiers are pumped to the sludge lagoons. The grit and paunch manure are stored in a diked earth basin adjacent to the plant.

In October 1968 the facilities for plant-scale chemical tests were completed. Multichemical systems as de-

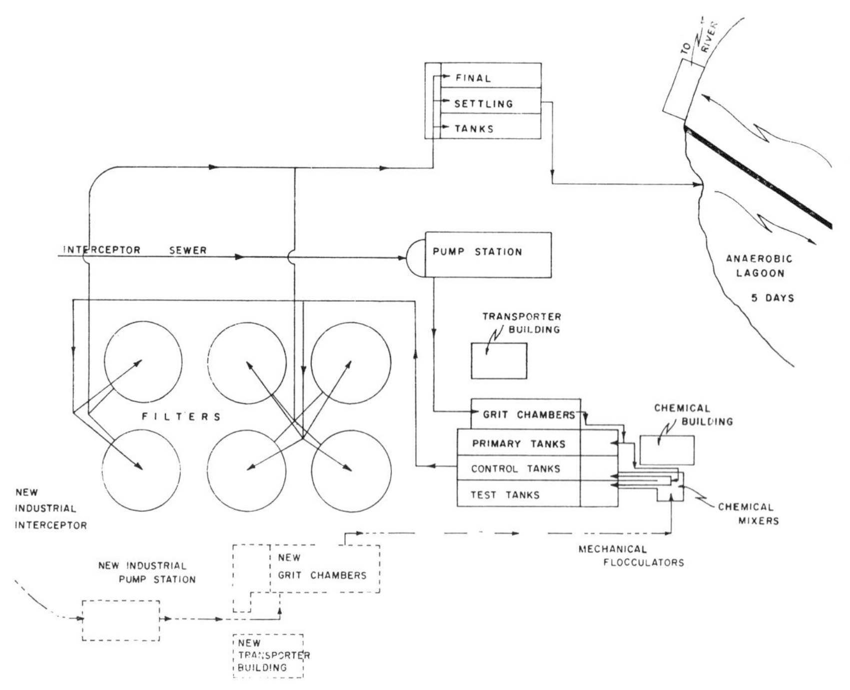

FIGURE 1.—Flow diagram of South St. Paul, Minn., wastewater treatment plant.

scribed in Patent Nos. 3,434,312 and 3,142,638 by Blaisdell and Klaas were investigated for several months without success. These multichemical systems consisted of bentonite clay, an anionic polyelectrolyte, ferric chloride, a cationic polyelectrolyte, and iron ore. After this study, laboratory investigations were made to determine if other chemical systems were effective on the combined waste.

Laboratory Evaluations

Laboratory evaluations were conducted on a six-gang floc stirrer with an adjustable paddle speed ranging from 0 to 120 rpm. One liter of waste was used on each stirrer. Initial investigations were based on qualitative visual observations on the premise that visual flocculation must occur before the chemicals could be considered active. This qualitative evaluation was conducted under three categories: (*a*) single-chemical systems, (*b*) dual-chemical systems, and (*c*) flocculation consistency with time. The chemicals were added to the raw degritted waste. After approximately 1 min of flash mixing at 120 rpm, the speed was reduced to approximately 60 rpm and was again reduced after a few minutes to 15 rpm. This speed of 15 rpm was intended to represent the hydraulic movements in the plant. Flocculation activity was based from the point of chemical addition to the point of sedimentation at 15 rpm.

Single-Flocculant Systems

This type of system would be the most practical and economical from the standpoint of handling and feeding the chemicals. Table I illustrates the evaluation of the single-flocculant systems. In all, 24 chemicals were tested for flocculation activity. These chemicals included anionic, cationic, and nonionic organics in addition to the inorganic salt, ferric chloride.

In summary of the work on cationics: Calgon ST-260 * and Hercules 220 † were effective in flocculation, but at high dosages. Calgon ST-260 formed a small, voluminous floc that would not be compatible with settling in the hydraulically overloaded South St. Paul plant. Hercules 220 formed a large, heavy floc that would be more desirable but would be extremely costly. The optimum dosages of these materials would be economically prohibitive, as is shown in Table II, which presents a summary of single-flocculant testing results.

The anionic materials found to be effective alone were Dow A-21 ‡ and Nalco 610.§ Further investigations were conducted on these two polymers because of their high activity at relatively low cost. A-1 flocculant formed a large, dense floc rapidly at an optimum dosage of approximately 4 mg/l. This was the type of flocculation that was thought necessary in order to combat the hydraulic overload condition of the plant. Later in the testing program the A-2 solution was determined to be economically prohibitive.

In addition to the organic flocculants tested, anhydrous $FeCl_3$# was tested and found to be active at an optimum dosage of approximately 100 mg/l. At this dosage the flocculation was not rapid and the ferric hydroxide floc was voluminous, which would indicate poor settling characteristics under dynamic conditions.

* Manufactured by Calgon Corp., Pittsburgh, Pa. Mention of products and manufacturers is for identification only and does not imply endorsement by EPA-OWP.

† Manufactured by Hercules, Inc., Wilmington, Del.

‡ Manufactured by Dow Chemical Co., Midland, Mich.

§ Manufactured by Nalco Chemical Co., Chicago, Ill.

Manufactured by Pennsalt Chemical Corp., Philadelphia, Pa.

TABLE I.—Single-Flocculant Evaluation

Chemical	Optimum or Maximum Dosage (mg/l)	Results
Calgon ST-260	20	Medium, voluminous floc formed
Calgon 269	2	Poor flocculation
Nalco 607	20	Poor flocculation
Nalco 675	2	Poor flocculation
Nalco 610	20	Medium floc with some clarity
Nalco 603	40	No activity
Hercules 220	20	Large, dense floc formed
Dow A-21	4	Large, dense floc formed
Dow A-22	2	No activity
Dow A-23	2	No activity
Dow C-31	50	No activity
Dow C-32	50	No activity
Dow SA 118.1A	50	No activity
Dow SA 11881.F	50	No activity
Dow SA 1569	50	No activity
Dow SA 1767	40	No activity
Dow 1621.2	40	No activity
Dow N-11	2	No activity
Dow N-12	2	No activity
Dow N-17	2	No activity
$FeCl_3$ (Pennsalt anhydrous)	100	Medium voluminous floc with excellent clarity
General Mills 162	35	No activity
General Mills 263	35	No activity
Polyethyleneimine (plant)	50	Fine voluminous floc formed

Dual-Flocculant Systems

The dual-flocculant systems were tested primarily to locate an anionic and cationic system that would work synergistically to reduce cost and increase treatment efficiency. The mechanism involved in this type of system is that the cationic materials normally form a small voluminous floc with extreme clarity; however, this floc is resistant to sedimentation under dynamic conditions. A high-molecular-weight anionic material is then added to increase the size and density of the floc significantly. Several cationic and anionic dual systems were tested. A

TABLE II.—Single-Flocculant Summary

Chemical	Price ($/lb)	Optimum Dosage (mg/l)	Treatment Cost ($/mil gal)
$FeCl_3$ (Pennsalt anhydrous)	0.055	100	46
Calgon ST-260	1.75	20	290
Hercules 220	1.75	20	290
Dow A-21	1.00	4	33
Nalco 610	2.45	2*	41*

* Later determined to be 20 and 410.

Note: Lb × 0.454 = kg; mil gal × 3,785 = 3,785 = cu m.

TABLE III.—Dual-Flocculant Evaluation

Chemical No. 1	Chemical No. 2	Results
$FeCl_3$ (Pennsalt anhydrous)	Dow A-21	No synergism
$FeCl_3$ (Pennsalt anhydrous)	Nalco 610*	No synergism*
$FeCl_3$ (Pennsalt anhydrous)	Nalco 675	Synergism suspected
Calgon 260	Dow A-21	No synergism
Hercules 220	Dow A-21	No synergism
Calgon 260	Nalco 610	No synergism
Hercules 220	Nalco 610	No synergism
$FeCl_3$ (Pennsalt anhydrous)	Iron ore	No synergism
$FeCl_3$ (Pennsalt anhydrous)	Calgon 260	No synergism
$FeCl_3$ (Pennsalt anhydrous)	Reten 220	No synergism
$FeCl_3$ (Pennsalt anhydrous)	A-22	No synergism
$FeCl_3$ (Pennsalt anhydrous)	A-23	Synergism suspected
Dow A-21	Iron ore	No synergism

* Synergism was suspected until an error of concentration was discovered.

list of these systems is given in Table III. The procedure for these evaluations was similar to that for the single-flocculant-system evaluations, except that after a rapid mix and flocculation with the cationic material at about 60 rpm, the speed of the stirrer was increased again to a maximum of 120 rpm and the anionic material was added to the waste. Anionic polymers were tested with ferric chloride. Synergistic activity was observed for ferric chloride with Nalco 675 and ferric chloride with Dow A-23. The dual system of ferric chloride and Nalco 610 was also initially encouraging.

Flocculation Consistency

Following the screening of the single- and dual-chemical systems available, a selection of the most active and economical systems was made for further laboratory evaluations. At this point three systems seemed feasible: (*a*) Dow A-21 alone, (*b*) dual system of Nalco 675 and $FeCl_3$, and (*c*) dual system of Dow A-23 and $FeCl_3$.

The purpose of the consistency tests was to observe the reaction of a chemical system with changes of the waste during the day. With these tests suspended solids (SS) analyses were conducted on samples withdrawn from 2 in. (5.1 cm) below the surface at paddle speeds of 15 rpm to obtain estimates of removal of SS and to compare the three systems under evaluation. The SS test was used as the guiding parameter because the chemicals were reacting directly with the SS. The results are shown in Table IV.

The SS test results in Table IV confirm the previous visual observations. The results also confirm the voluminous nature of the ferric hydroxide floc and its resistance to settling under dynamic conditions. If the ferric-chloride-treated waste had been allowed to settle under static conditions, it would have shown the best overall results.

The laboratory evaluations of the chemicals for the treatment of the combined waste are:

1. Dow A-21 was the most active, economical, and consistent of the single-chemical systems tested at dosages between 2 and 5 mg/l.
2. There were definite indications of synergism and increased performance over a single system in both dual systems, that is, $FeCl_3$ plus Nalco 675 and $FeCl_3$ plus Dow A-23.
3. These systems consistently flocculated during diurnal changes in waste composition.

Plant-Scale Evaluations

After the laboratory work, three systems were planned to be evaluated on a plant-scale basis: (*a*) the single chemical, Dow A-21, at a dosage range of 2 to 5 mg/l, (*b*) the dual systems of 20 to 50 mg/l $FeCl_3$ plus 0.75 to 1.25 mg/l Dow A-23 and Nalco 675, and (*c*) the inorganic system of $FeCl_3$ alone at a dosage range of up to 100 mg/l. The chemicals were to be added to the raw degritted waste before entering the flocculation tank. In parallel with the test tank were control tanks to which no chemicals were added. These two systems were previously checked to insure the duplicability of treatment. They were essentially equal in treat-

TABLE IV.—Laboratory Results on Consistency Tests

Day	Time	Chemical	Dosage (mg/l)	Suspended Solids (mg/l)		
				Degritted Raw	Control	Treated
6/19	6:00 AM	Dow A-21	2.0	192	142	53
6/19	6:00 AM	$FeCl_3$*/Nalco 610	10/10	192	164	47
6/19	6:00 AM	$FeCl_3$*/Nalco 610	20/10	192	164	49
6/19	9:00 AM	Dow A-21	3.0	1,111	769	560
6/19	9:00 AM	Dow A-21	4.0	1,111	769	311
6/19	9:00 AM	Dow A-21	5.0	1,111	769	204
6/19	9:00 AM	$FeCl_3$*/Nalco 610	40/10	1,111	824	144
6/19	9:00 AM	$FeCl_3$*/Nalco 610	50/10	1,111	824	172
6/19	9:00 AM	$FeCl_3$*	50	1,111	824	1,051
6/20	9:30 AM	Dow A-21	4.0	804	704	268
6/20	9:30 AM	Nalco 610	2.5	804	692	824
6/20	9:30 AM	$FeCl_3$*/Nalco 610	50/1	804	728	884
6/20	10:40 AM	Dow A-21	4.0	1,040	956	292
6/20	10:40 AM	$FeCl_3$*/Dow A-21	40/3	1,040	976	1,028
6/20	10:40 AM	$FeCl_3$*/Dow A-21	40/4	1,040	976	580
6/20	10:40 AM	$FeCl_3$*	100	1,040	976	1,276
6/20	12:40 PM	Dow A-21	4.0	1,200	1,136	312
6/20	12:40 PM	Dow A-21	5.0	1,200	1,136	252
6/20	12:40 PM	$FeCl_3$*/Nalco 610	40/1	1,200	1,064	1,220
6/20	12:40 PM	$FeCl_3$*/Nalco 610	40/5	1,200	1,064	696
6/20	12:40 PM	$FeCl_3$*/Nalco 610	40/10	1,200	1,064	140
6/25	10:30 AM	$FeCl_3$*/Dow A-23	30/1.5		580	108
6/25	10:30 AM	$FeCl_3$*/Dow A-23	40/1.5		580	80
6/25	10:30 AM	$FeCl_3$*/Nalcolyte 675	40/1.5		588	100
6/25	10:30 AM	$FeCl_3$*/Nalcolyte 675	50/1.5		588	80
6/25	3:00 PM	$FeCl_3$*/Nalcolyte 675	40/1.5		556	132
6/25	3:00 PM	$FeCl_3$*/Nalcolyte 675	50/1.5		556	104
6/25	3:00 PM	$FeCl_3$*/Dow A-23	10/1.5		556	176
6/25	3:00 PM	$FeCl_3$*/Dow A-23	20/1.5		556	164
6/25	3:00 PM	$FeCl_3$*/Dow A-23	50/1.5		556	44
6/26	12:45 PM	$FeCl_3$*/Nalcolyte 675	30/1.5		900	264
6/26	12:45 PM	$FeCl_3$*/Nalcolyte 675	40/1.5		900	216
6/26	12:45 PM	$FeCl_3$*/Nalcolyte 675	50/1.5		900	184
6/26	1:45 PM	$FeCl_3$*/Dow A-23	20/1.5		728	248
6/26	1:45 PM	$FeCl_3$*/Dow A-23	30/1.5		728	144
6/26	1:45 PM	$FeCl_3$*/Dow A-23	40/1.5		728	116
6/27	7:20 AM	$FeCl_3$*/Nalcolyte 675	10/1.5		284	56
6/27	7:20 AM	$FeCl_3$*/Nalcolyte 675	20/1.5		284	32
6/27	7:20 AM	$FeCl_3$*/Dow A-23	10/1.5		284	60
6/27	7:20 AM	$FeCl_3$*/Dow A-23	20/1.5		284	24

* Pennsalt anhydrous.

ment. Again, as in the laboratory evaluations, SS was the primary parameter used to evaluate performance with BOD analyses periodically.

Single Chemical Systems

Figure 2 illustrates the treatment results using Dow A-21 in comparison to a control. The dosage of Dow A-21

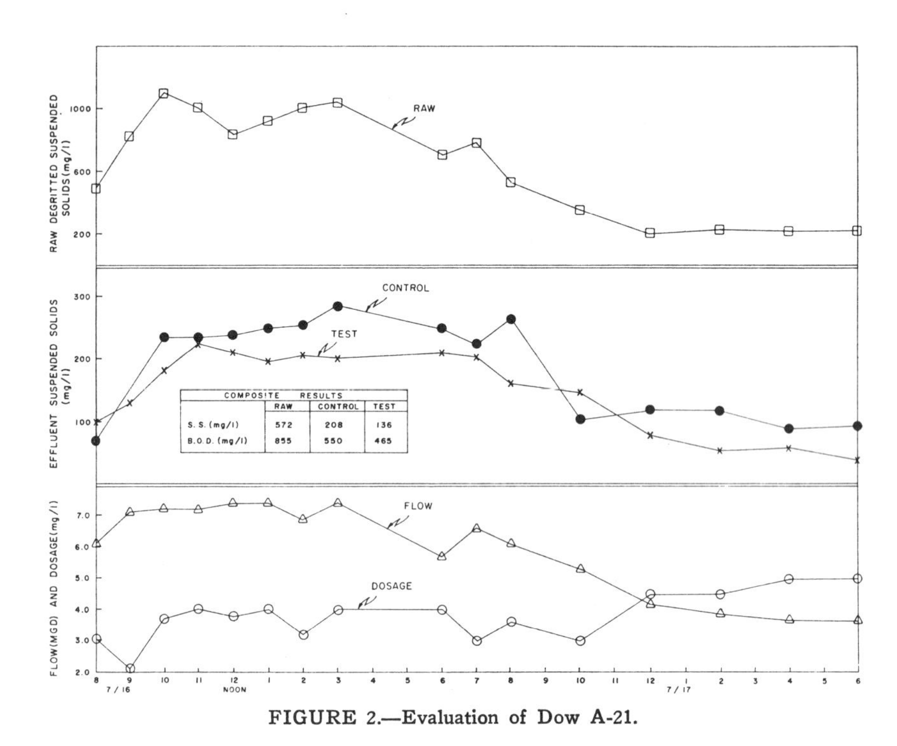

COMPOSITE RESULTS	RAW	CONTROL	TEST
S. S. (mg/l)	572	208	136
B.O.D. (mg/l)	855	550	465

FIGURE 2.—Evaluation of Dow A-21.

ranged from 2 to 5 mg/l during the testing period. This dosage range gave an average SS in the test effluent of 136 mg/l versus a control effluent of 208 mg/l, with the raw degritted waste containing 572 mg/l. After a short period of testing there was visual evidence that the results were not as satisfactory as anticipated. The chemical formed a massive floc within seconds after addition to the waste. It was learned that this floc must be broken up and re-formed before sedimentation would take place; otherwise extreme flotation occurred in the sedimentation tanks. This seems to be characteristic of the material. The manufacturer reported that the supply of the chemical would be limited. Therefore, no further plant testing work was undertaken on the single system.

Dual Chemical Systems

Laboratory work indicated that maximum efficiency of the dual system could be achieved by allowing the $FeCl_3$ to react with the waste for 10 to 15 min before the addition of the anionic polymers. This was achieved in the plant by introducing the anionic polyelectrolyte through a pipe drilled with holes and extending the width of the flocculation tank. A dye study indicated where the pipe should be installed to achieve a maximum of 15 min for primary flocculation.

In an effort to reduce the costs of the dual-chemical systems, other $FeCl_3$-containing materials were tested in the laboratory. A waste $FeCl_3$ material generated by the Buckbee Mears Manufacturing Company in St. Paul demonstrated activity similar to that of the commercial $FeCl_3$ at approximately one-sixth the possible cost. Naturally, the next step was to evaluate the waste $FeCl_3$ in the dual system.

Four dual systems were investigated. The cationic portion of the dual system

TABLE V.—Mean Dosages of Chemicals

Test	Chemical Addition	Dosage (mg/l)	
		8 AM to 10 PM	10 PM to 8 AM
I	Nalco 675 Pennsalt anhydrous $FeCl_3$	0.97 44.9	0.77 28.1
II	Nalco 675 Waste $FeCl_3$	0.97 46.3	0.75 38.6
III	Dow A-23 Pennsalt anhydrous $FeCl_3$	0.91 36.8	0.77 33.3
IV	Dow A-23 Waste $FeCl_3$	0.78 42.9	0.70 27.3

was ferric chloride in both the anhydrous and waste forms. The anionic polymers were Nalco 675 and Dow A-23. The four combinations of these cations followed by anionic polymers were tested. The dosages selected were 50 mg/l of $FeCl_3$ plus 1.0 mg/l of anionic polymer from 8 AM to 10 PM and 30 mg/l of $FeCl_3$ plus 0.75 mg/l of anionic polymer from 10 PM to 8 AM. The two time periods divide the waste into combined and domestic flows. The actual dosage varied, especially in Test III, when the $FeCl_3$ feed line was plugged for 5 hr. The mean dosages are shown in Table V.

In all four investigations, the test tank with the dual chemical additions performed better than the control tank without chemical addition. Figures 3 through 6 illustrate the SS concentrations in Tests I through IV, respectively.

These four sets of frequency curves indicate greater removals with the Nalco 675 anionic polymer. A change in removal because of two sources of ferric chlorides is not noticeable. The waste $FeCl_3$ has considerable economic advantages and no known technical disadvantages.

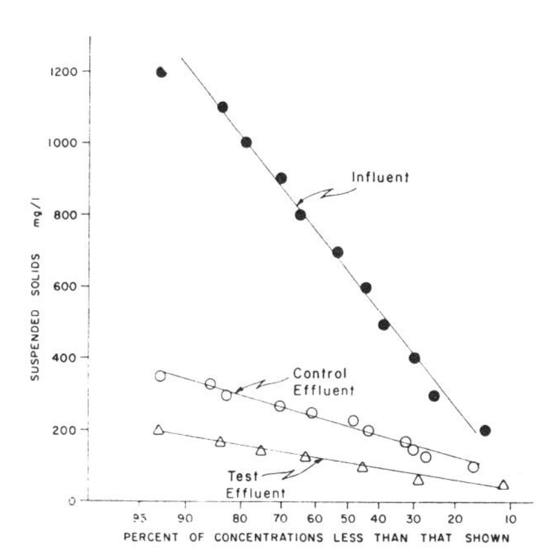

FIGURE 3.—Results obtained from anhydrous ferric chloride plus Nalco 675.

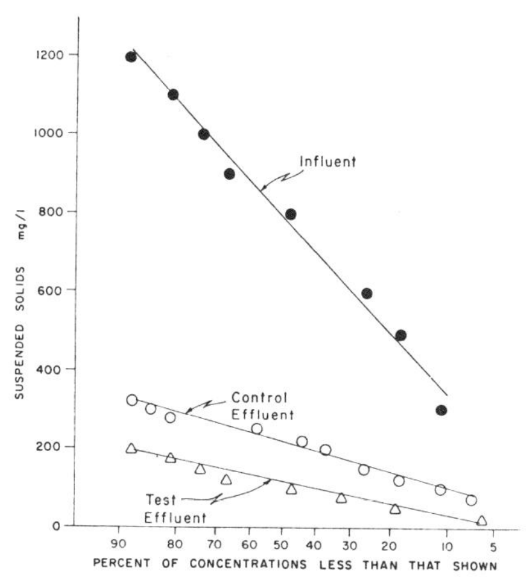

FIGURE 4.—Results obtained with waste ferric chloride plus Nalco 675.

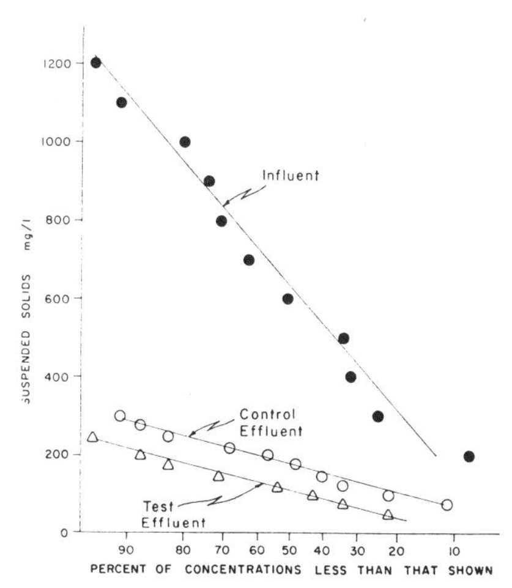

FIGURE 5.—Results obtained with anhydrous ferric chloride plus Dow A-23.

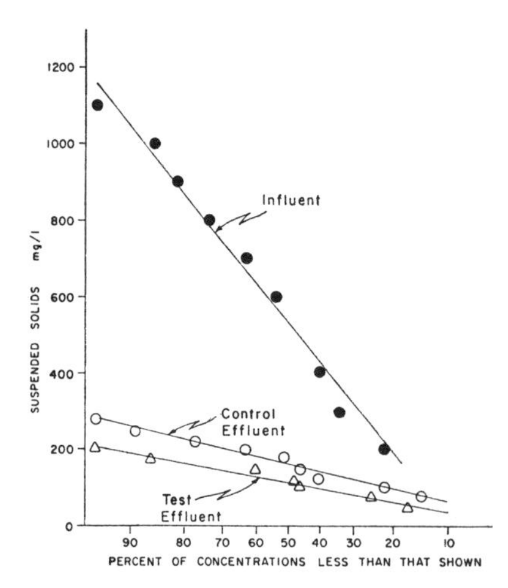

FIGURE 6.—Results obtained with waste ferric chloride plus Dow A-23.

Two methods of sampling were used. Grab samples were taken routinely through the day and analyzed for SS and, occasionally, for BOD. Samples were also taken routinely throughout the day and composited in proportion to flow. These samples were analyzed for SS and BOD. These data are summarized in Table VI. The data show the magnitude of concentrations from which removals can be calculated and compared. Two sampling schemes gave similar concentrations, thereby verifying the test results. The BOD remaining is mainly in a dissolved form.

Statistical hypothesis tests were made on the SS grab sample data. Using the "Student t" statistic, a significant difference was found at the < 0.01 probability level between test and control tanks in all four investigations. The odds are less than 1 in 100 that the difference was caused by chance alone. Statistical comparisons of the mean SS of different investigations were also made. The means were not significantly different between Dow A-23 plus anhydrous $FeCl_3$ and Nalco 675 plus anhydrous $FeCl_3$. The mean values between investigations using anhydrous $FeCl_3$ and waste $FeCl_3$ were not significantly different.

Single Chemical—$FeCl_3$

Flocculation of the waste with anhydrous $FeCl_3$ alone was attempted to demonstrate the increased efficiency of the dual system in comparison. The dosages were 50 and 100 mg/l. The $FeCl_3$ formed a small voluminous floc, as previously observed in the laboratory. The floc with no organic polyelectrolyte to increase the size and density was carried through the tank and over the weirs. This floc was readily observable in the primary effluent as it flowed over the weirs. The data illustrated in Figures 7 and 8 confirm the increase in SS in the effluent when $FeCl_3$ was added to the waste.

TABLE VI.—Summary of Dual System Tests

Sample Type	Chemical Addition	Mean SS (mg/l)			Mean BOD (mg/l)		
		Influent	Test	Control	Influent	Test	Control
Grab	A-23 + $FeCl_3$ Pennsalt anhydrous	621	113	174	1,322	549	663
	A-23 + $FeCl_3$ waste	552	117	162	1,244	604	707
	675 + $FeCl_3$ Pennsalt anhydrous	644	120	211	1,338	555	691
	675 + $FeCl_3$ waste	803	106	202	1,272	632	772
Composite	A-23 + $FeCl_3$ Pennsalt anhydrous	631	106	171	985	451	536
	A-23 + $FeCl_3$ waste	525	131	147	814	514	584
	675 + $FeCl_3$ Pennsalt anhydrous	635	130	183	952	486	618
	675 + $FeCl_3$ waste	875	118	220	1,178	519	584

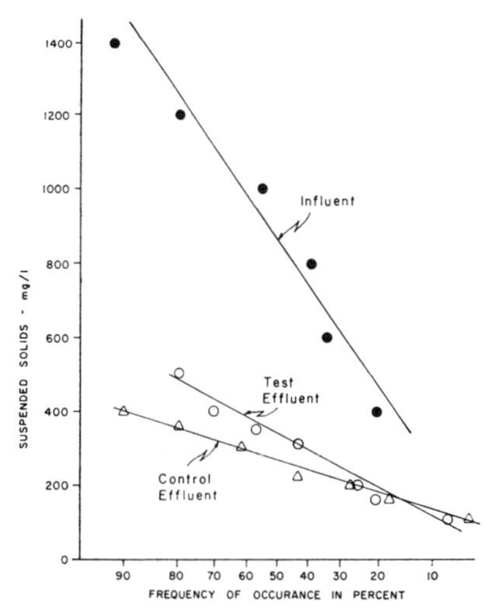

FIGURE 7.—Results obtained with 50 mg/l ferric chloride.

The addition of $FeCl_3$ alone was unsatisfactory.

Economics

Costs are calculated on the basis of a 10-mgd (37,850-cu m/day) flow. Labor, capital construction, and utility costs are based on information gained during the plant-scale evaluation. Chemical costs are based on dosage utilized in the plant investigation and price quotes from the chemical companies. The plant evaluation made with full-scale equipment offers reliable cost information. A summary of these costs is shown in Table VII.

The cost of anhydrous ferric chloride was figured at $0.055/lb ($0.12/kg), which is the price for large quantities delivered to the nearby Minneapolis–St. Paul plant. Both polyelectrolytes were priced at $1.60/lb ($3.50/kg) plus $0.02/lb ($0.04/kg) freight for 5,000-lb (2,270-kg) lots. The dosages were 50 mg/l $FeCl_3$ and 1 mg/l polyelectrolyte for 14 hr and 30 mg/l and 0.75 mg/l, respectively, for 10 hr. Handling costs were estimated at $3,500/yr.

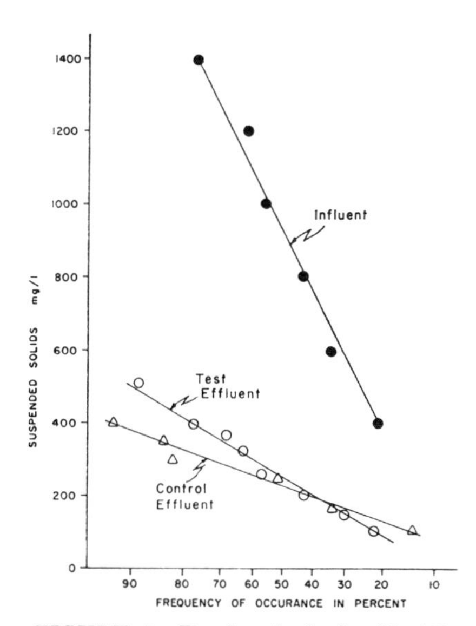

FIGURE 8.—Results obtained with 100 mg/l ferric chloride.

The capital construction was amortized over 30 yr for the chemical building, 5 yr for the feeding equipment, and 20 yr for the special piping, all at a 5 percent rate of interest. Maintenance of these items was estimated at $2,500/yr.

Labor costs were figured at 3,120 hr/yr at $5.00/hr. This manpower was necessary to operate the feeding equip-

TABLE VII.—Economics Based on a 10-Mgd Flow

Item	Cost	
	$/yr	% of Total
Chemical and handling cost	117,416	71
Capital construction and maintenance (piping, chemical building, and feeding equipment)	24,958	15
Labor (operator, sampler, chemist)	19,100	12
Utilities (gas, water, electricity)	2,760	2
Total	164,234	100

Note: 10 mgd = 37,850 cu m.

ment, prepare chemicals, and sample. Laboratory equipment and 0.3 man-years of a chemist's time cost an additional $3,500.

The costs of electricity, gas and water were on a monthly basis of $168, $20, and $42, respectively.

The total cost is $45/mil gal ($12/1,000 cu m).

Summary and Conclusions

The use of chemical flocculation has been demonstrated in the treatment of combined meat-packing and domestic wastes. Various chemical treatment schemes were demonstrated in an overloaded primary portion of a trickling filter plant in St. Paul, Minn., after intensive laboratory evaluations.

Of the various treatment schemes demonstrated, a dual system of ferric chloride, followed by the addition of a high-molecular-weight anionic organic polyelectrolyte, proved to be the most successful. In round numbers this system decreased SS concentrations by 100 mg/l and BOD concentrations by 140 mg/l, leaving effluent concentrations of 120 mg/l and 550 mg/l, respectively. The primary flocculation of the colloidal and fine particles by the ferric chloride produced a small voluminous floc that was further increased in size and density by the secondary flocculation of the organic polyelectrolyte, producing a settleable precipitate. The ferric chloride alone was unsatisfactory because it produced a nonsettleable floc under prevailing hydraulic conditions.

This investigation also demonstrated the value of correlative jar testing before field testing of flocculants. Laboratory jar testing was used to screen the various chemical treatment schemes to select the best for field demonstrations. This approach proved to be valuable in the savings of time and money.

The costs of full-scale application of the dual-chemical system of anhydrous $FeCl_3$ and Dow A-23 or Nalco 675 are 71 percent for chemicals, 15 percent for construction, 12 percent for labor, and 2 percent for utilities.

The cost of chemically treating combined 90 percent meat-packing and 10 percent domestic wastes would be approximately $0.045/1,000 gal ($0.012/cu m) of waste treated. As the strength of the waste decreases, the chemical costs will be reduced. These costs are in addition to the normal operating costs of the existing treatment facilities.

ENFORCEMENT IN WATER POLLUTION CONTROL

Murray Stein; Thomas C. McMahon; Allen S. Lavin, Frederick M. Feldman, Sanford R. Gail; H. Edward Dunkelberger, Jr.

FEDERAL VIEWPOINT

Murray Stein

Since its founding in 1928, the Federation has contributed much to the advancement of fundamental and practical knowledge of water quality management. Its Annual Conference is an influential gathering of experts and decision-makers in the profession, government, industry, and other callings, who share with one another and with their fellow citizens a profound concern about the deterioration of water, air, and land resources, and a resolve to work for the restoration of the natural environment.

Environmental Protection Agency

With the assent of Congress to the President's Reorganization Plan No. 3 of 1970, the Environmental Protection Agency (EPA) will come into being on December 2, 1970. The functions of the Federal Water Quality Administration (FWQA), U. S. Department of the Interior, will be transferred to the new agency. The functions of the Secretary of the Interior under the Federal Water Pollution Control Act and related authorities will be transferred to the Administrator of EPA. Air pollution, solid wastes, pesticides (in part), radiation (in part), and other environmental problem areas will be part of the broad mandate of EPA. The reorganization, in the President's words, will "give unified direction to our war on pollution and provide a stronger organizational base for our stepped-up effort."

Development of Federal Water Quality Law

Under the commerce clause of the United States Constitution, Congress exercises jurisdiction over waters capable of use as highways of interstate or foreign commerce, whether interstate or intrastate. Congress has determined, as a matter of policy, the extent to which that jurisdiction would be applied in various aspects of water resources management. Despite a clear constitutional basis and a growing need, the federal role in water quality control had hesitant beginnings.

Refuse Act

The River and Harbor Act of 1899, in Section 13, known as the Refuse Act, prohibited the discharge or deposit, without a permit, of refuse other than liquid wastes flowing from streets and sewers into navigable waters. That statute has been administered through the years by the Secretary of the Army (formerly the Secretary of War), acting through the Chief of Engineers, primarily in the interest of navigation. It has been held by the courts not to be limited to refuse matter that impedes or obstructs navigation, and it has been applied, particularly in recent months through consultation and coordination among the departments of Justice, Army, and the Interior, as an additional water pollution abatement tool.

The 1899 law is the basis of the 1970 federal suits to halt mercury discharges into navigable waters by 8 industries at 10 locations in the U. S. The viability of the Refuse Act is evidenced by the plans of the Department of the Army to require permits for future discharges into navigable waters and to require that applications for such permits be accompanied by a state certification under Section 21(b) of the Federal Water Pollution Control Act, added in 1970, with respect to compliance with applicable water quality standards. Further, the Department of the Army intends to require that a discharger disclose its discharges in order to recive a permit.

Public Health Service Act

The Public Health Service Act of 1912 authorized investigations of water pollution relating to the diseases and impairments of man. With the transfer to the new EPA of the functions of the Bureau of Water Hygiene [Environmental Control Administration, U. S. Public Health Service (USPHS), Department of Health, Education, and Welfare], the traditional concerns of water and human health will be related more directly to total environmental protection.

Oil Pollution Abatement

The Oil Pollution Act, 1924, was designed to control oil discharges from vessels into coastal navigable waters. It was not until 1966 that its coverage was extended to inland navigable waters. The same enactment transferred its administration from the Secretary of the Army to the Secretary of the Interior and made its expanded provisions very difficult to enforce because of the requirement that an oil discharge must be grossly negligent or willful before a penalty or cleanup liability would apply. The Water Quality Improvement Act of 1970 repealed the 1924 Act and added new and strengthened provisions for the control of oil pollution to the basic Federal Water Pollution Control Act (Section 11).

Under the Act, and the President's delegations of authority and the National Contingency Plan issued thereunder, the FWQA is responsible for enforcement activities related to: (*a*) violations in inland waters of regulations on harmful discharges of oil and methods and procedures for oil removal; (*b*) violations of regulations for the prevention of discharges from onshore and offshore facilities, except facilities related to transportation; and (*c*) the abatement of threatened oil discharges under the authority to seek injunctive relief in cases of imminent or substantial threat to the public health or welfare because of an actual or threatened oil discharge from an onshore or offshore facility. The Act prohibits oil discharges into navigable waters except as permitted by regulation. FWQA regulations effective on their publication in the *Federal Register* on September 11, 1970, provide that prohibited harmful discharges include those that violate applicable water quality standards or those that cause a film or sheen on, or discoloration of, the surface of the water or adjoining shorelines, or cause a sludge or emulsion to be deposited beneath the surface of the water or on adjoining shoreline. Discharges from a properly functioning vessel engine are excepted, but oil accumulated in a vessel's bilges is not. Demonstration projects relating to oil pollution control may be permitted.

The oil pollution section extends the Federal Water Pollution Control Act for the first time to the waters of the Contiguous Zone, the 9-mile (14.4-km) zone beyond the territorial sea provided by international convention. It provides a criminal penalty only for the failure to notify the appropriate federal agency (U. S. Coast Guard) immediately of a prohibited discharge. Civil penalties are provided for knowingly discharging oil in harmful quantities and for violation of regulations. The section establishes the principle of absolute liability for the assumption of oil removal costs by the owner or operator of a vessel, an onshore facility, or an offshore facility. Proof that a discharge was caused by an act of God, an act of war, negligence on the part of the U. S., or an act or omission of a third party whether negligent or not, is required to avoid liability. Limits of liability are set, but proof of willful negligence or willful misconduct within the privity and knowledge of the owner makes the owner liable for the removal costs without limitation.

Comprehensive Water Quality Law

This discussion of navigation, health, and oil pollution statutes has brought us beyond the point in time at which Congress finally gave statutory definition to the federal role and responsibility in water pollution control. Many bills were introduced over the years to establish a comprehensive water pollution control program. On three occasions legislation neared enactment. Success was delayed until 1948 and the passage of Public Law 845, 80th Congress, which authorized a 5-yr program. The law was extended for another 3 yr by Public Law 579, 82nd Congress, to June 30, 1956.

The Federal Water Pollution Control Act became permanent law in 1956, with the enactment of Public Law 660, 84th Congress. The Act has been strengthened by (*a*) the Federal Water Pollution Control Act Amendments of 1961 (Public Law 87–88), which among other provisions, extended the federal enforcement authority to navigable as well as interstate waters; (*b*) the Water Quality Act of 1965 (Public Law 89–234), which among other provisions created the Federal Water Pollution Control Administration (now FWQA) and authorized the establishment of water quality standards for the nation's interstate waters; (*c*) the Clean Water Restoration Act of 1966 (Public Law 89–753), which among other provisions authorized a vastly increased level of federal assistance for the construction of municipal wastewater treatment facilities; and (*d*) the Water Quality Improvement Act of 1970 (Public Law 91–224), which strengthened the law in the areas of wastewater from watercraft, mine drainage, lake eutrophication, pollution problems in the Great Lakes, manpower needs in water quality management, pesticides, pollution by activities under federal license or permit, and pollution by oil. Limited authority is provided for the control of pollution by hazardous polluting substances other than oil, with a study and report directed to the need for and desirability of legislation to impose liability for the cost of removal of hazardous substances, with attention to enforcement, including the imposition of penalties and recovery of costs if removal is undertaken by the government. Regulations designating hazardous substances for purposes of the law are in preparation.

Enforcement and Water Quality Standards

Policy

Water quality management is an exciting case study in intergovernmental relations. The Federal Water Pollution Control Act declares it to be the policy of Congress to recognize, preserve, and protect the primary responsibilities and rights of the states in preventing and controlling water pollution, and it expressly states that no right or jurisdiction of the states with respect to their waters is impaired or affected thereby. A statement of the Act's purpose to enhance the quality and value of our water resources and to establish a national policy for the prevention, control, and abatement of water pollution was added to the law in 1965. The multiple activities of FWQA are conducted under the letter and in the spirit of that declaration of policy.

Authorities

The Federal Water Pollution Control Act has provided the statutory base for a wide range of water quality management activities—basin planning; basin planning grants; research, investigations, and demonstrations, and for these purposes grants, contracts, fellowships, and training; investigations of specific problems confronting public entities and industrial plants; basic water quality data collection and dissemination; demonstration grants in the areas of combined and storm sewer overflow, advanced waste treatment and water purification and joint municipal-industrial waste treatment, and industrial pollution control; state program grants; and municipal wastewater treatment plant construction grants.

The Act authorizes two distinct but related regulatory activities. One is the abatement of pollution of interstate or navigable waters that endangers the health or welfare of persons. The other is the establishment of water quality standards for interstate waters, including coastal waters, the federal authority to be invoked in the first instance only if a state fails to adopt acceptable criteria and a plan for their implementation and enforcement for its interstate waters; revision of the standards; and the abatement of discharges that lower water quality below the standards.

Three-Stage Enforcement Procedure

Fifty-one actions have been taken to abate pollution that is endangering the health or welfare of persons. This involves a three-stage procedure—conference, public hearing, and court action—each succeeding stage to be reached only if satisfactory progress to abate pollution is not made at the preceding stage. The conference is a nonadversary proceeding. Conferees are the state water pollution control agencies, any interstate water pollution control agency, and the Department of the Interior. The conference inquiries into the occurrence of pollution subject to abatement under the Act, the adequacy of measures taken toward abatement, and the nature of any delays. If possible, the conferees reach agreement on a required program of remedial action. The public hearing is a formal procedure directed toward individual alleged polluters. The findings and recommendations of the Hearing Board, which is comprised of five or more members appointed by the Secretary of the Interior Administrator, are sent to the polluters with a notice specifying a time for compliance and to the states and any appropriate interstate agencies. Court action is the ultimate resort. The court has jurisdiction to enter such judgment, and orders to enforce it, as the public interest and the equities of the case may require.

The authority to abate pollution that is endangering the health or welfare of persons may be invoked on federal initiative on the basis of reports, surveys, or studies, or at state request. The request of the Governor is a prerequisite to enforcement action in the case of the intrastate pollution of interstate or navigable waters. The enforcement authority may be invoked on federal initiative to abate pollution, interstate or intrastate, that impairs the interstate marketing of shellfish or shellfish products. International pollution situations are subject to abatement on the basis of reports, surveys, or studies from an international agency and the request of the U. S. Secretary of State.

Of the 51 actions taken under this enforcement procedure, 30 have been taken on federal initiative, 15 at state request, and 6 on both bases. The shellfish provision has been invoked in five cases, including the Boston (Mass.) Harbor conference. Only four cases have reached the public hearing stage. and only one case has been the subject of court action.

The four-state Lake Michigan conference, which met in the fall of 1970 in workshop sessions directed particularly to the discussion of valid temperature requirements for the lake, is a good example of an enforcement conference that has served as a forum and a catalyst for the identification and the solution of complex pollution problems in a heavily populated, highly industrialized area.

Water Quality Standards

The law directs that water quality standards for the nation's interstate and coastal waters be such as to protect the public health or welfare, enhance the quality of water, and serve the purposes of the Federal Water Pollution Control Act. With the establishment of water quality standards, for the first time in water areas throughout the nation there exists a specified set of conditions to which to adhere and to seek in enhancing and protecting water quality.

Each of the 50 states and 4 other jurisdictions met the October 2, 1966, statutory deadline for filing a letter of intent to adopt water quality criteria applicable to its interstate waters and a plan for the implementation and enforcement of the criteria. The criteria—scientific requirements on which judgments may be based as to the suitability of water quality to support a designated use—and the plan, taken together, comprise the water quality standards. Each of the states, including the four other jurisdictions, met the June 30, 1967, deadline for the establishment of the standards. As of September 30, 1970, the standards of 22 states had been fully approved, including an antidegradation statement, committing a state to the protection of its high quality waters. The standards of eight states had been approved fully, not including the antidegradation statement, those of 15 states approved in part with the antidegradation statement and those of nine states approved in part without the antidegradation statement. A major program thrust at the present time, involving in some cases intensive negotiations with the states, is the resolution of the remaining exceptions to the water quality standards, including the antidegradation statements. Federal approval has been withheld in part from standards submitted by the states for various reasons. Among them are inadequate temperature criteria or dissolved oxygen (DO) criteria for the protection of present or future water uses, inadequate implementation plans in that the waste treatment requirements are too low, and time schedules too lenient for the construction of necessary abatement facilities.

An important part of FWQA's responsibility is to work with the states in the refinement of criteria and the revision of water quality standards in response to the law's directive that standards shall enhance the quality of water.

Another aspect of the federal effort is to remain currently informed of the status of the actions taken by industrial, municipal, and other dischargers to implement the criteria. For the purpose of assessing compliance with the treatment requirements and time schedules in the implementation plans, which are an integral part of the standards, a broad monitoring network has been developed in cooperation with the states, using the resources of the U. S. Geological Survey to the extent possible. A computerized capability will be developed to obtain at any time a full status report on standards compliance.

Federal Establishment of Water Quality Standards

The law authorizes the Secretary of the Interior to act to establish water quality standards for the interstate waters of a state if it fails to take timely and acceptable action. This was done for the first time in the matter of certain aspects of the water quality standards of the state of Iowa, which were accepted in part as federal standards on January 16, 1969. Secondary treatment of wastes discharged to the Mississippi River, temperature criteria applicable to interstate waters except the boundary rivers, and matters relating to dilution, phenols, public water supply quality, and disinfection remained at issue between the U. S. Department of the Interior and the state after a conference held in April 1969 and exhaustive discussions.

The Secretary's proposed regulations setting forth water quality standards applicable to interstate waters of Iowa were published in the *Federal Register* on November 1, 1969. The Secretary adopted the regulations, which would have taken effect 30 days after publication in the *Federal Register* on May 12, 1970, if the state had not petitioned for a hearing before a hearing board empowered to make findings as to whether the standards should be approved or modified. The state of Iowa has availed itself of this statutory right. Arrangements are under way to convene a hearing board of five or more persons to hold a public hearing in the area affected by the standards. The state, the U. S. Department of Commerce, the U. S. Department of Health, Education, and Welfare, and other affected federal agencies, may each select a member of the hearing board. Officers or employees of the U. S. Department of the Interior must be in the minority of the hearing board membership.

FWQA is working with the state of Iowa to limit the matters to be considered by the hearing board to those still at issue. Former differences have been resolved successfully, and the key question before the hearing board will be that of secondary treatment of municipal wastes discharged by one community into the Mississippi River. The Secretary's regulations call for a minimum of secondary treatment of municipal wastes and equivalent treatment of industrial wastes discharged into the boundary rivers by December 31, 1973. Treatment requirements for the interior interstate streams are not at issue. It is the FWQA's position that water quality enhancement demands a higher degree of treatment than heretofore required and that under present law, the implementation plan is the key to the timely achievement of adequate treatment through the water quality standards.

In the case of the water quality standards of another state, Virginia, the Secretary had called a conference late in 1969 to consider matters at issue between the state and the Department. He subsequently postponed the conference when further discussions with Virginia authorities gave promise of agreement on water use designations, water quality criteria, and waste treatment requirements and implementation plans that had been found unacceptable in part. Both the letter and the spirit of the Water Quality Act and the basic Federal Water Pollution Control Act encourage the assumption by the states of the primary responsibility for the protection and enhancement of the quality of their waters.

Enforcement of Water Quality Standards

The law makes subject to abatement the discharge of matter into interstate waters that reduces their quality below the water quality standards established pursuant to the Water Quality Act. Unlike the enforcement procedure in the case of pollution of interstate or navigable waters that endangers the health or welfare of persons, the abatement of water quality standards violations is reached through direct court action. Under present law the consent of a Governor is required, as it is in court actions arising under the older enforcement procedure in the case of violations that are intrastate in effect.

At least 180 days before an abatement action is initiated, the Secretary of the Interior must notify the violators and others interested of the violation of the standards. The legislative history of the law directs that during this period an informal hearing shall be held on the request of a state, an alleged violator, or another interested party. Regulations provide for the conduct of such a hearing as well as for the conduct of the standards-setting conference and the public hearing for which the law provides.

The first notices of violation were issued in 1969 in six cases. These involved a mining company in Kansas, and four steel companies and a city in the Lake Erie Basin. Informal hearings were held in these cases and, as expected, court action against the dischargers is being avoided through voluntary compliance. The Secretary issued five more 180-day notices in 1970 to alleged violators of water quality standards. These cases involve a railroad company in New York State, a chemical company in New Jersey, a feed company in Nebraska, a drainage district in Kansas, and a city in North Dakota. Informal hearings were held in four of these cases and the fifth was postponed. Hopefully, voluntary compliance by the dischargers will make court action unnecessary.

Most of the water quality standards violation notices issued thus far have been directed to dischargers in water areas already involved in enforcement actions taken under the older abatement route. More and more FWQA enforcement and regulatory activity will be focused on water quality standards implementation and compliance. Administration legislative proposals would integrate more directly the standards compliance—enforcement conference mechanisms.

In the enforcement of water quality standards, as in their establishment, the states have a first responsibility. Where and when the states do not take prompt, effective action, the federal authority will be applied to protect the quality of the nation's water resources.

Administration Legislative Proposals

The President, in his landmark message of February 10, 1970, to Congress on the environment, set forth a 37-point program for the restoration of the natural environment. Among the 14 points addressed to the control of water pollution were proposals for far-reaching changes in the water quality standards and enforcement provisions of the Federal Water Pollution Control Act.

In summary, the Administration bill to amend Section 10 and certain other sections of the Act (S. 3471, H. R. 15905, H. R. 16036, and other bills) would expressly apply the Act's abatement provisions to boundary water as well as interstate and navigable water, the tributaries of these waters, groundwater, water of the Contiguous Zone and, under certain circumstances, the high sea. Water quality standards, which now consist of water quality criteria and a plan for their implementation and enforcement, would include three elements—water quality criteria; water quality requirements controlling discharges affecting water quality, that is, effluent requirements; and the implementation and enforcement plan.

Standards, established for interstate waters under present law, would be established for other waters covered by the amended law. The Secretary would act initially with respect to water under state jurisdiction only in the absence of satisfactory state action. He would establish standards for water not under state jurisdiction. The authority to abate discharges that violate the standards is made directly applicable to discharges that reduce water quality below established standards, discharges of lesser quality than discharge requirements, and discharges not in compliance with the implementation and enforcement plan. The consent of a Governor would no longer be required in cases of intrastate standards violations nor in cases of post-conference court action involving intrastate pollution. The court would be directed to include enforcement conference recommendations in activities enjoined. Comparable provisions would integrate water quality standards into enforcement actions taken through the conference route and into any post-conference court actions. Effective 6 months after enactment, the court could impose a new penalty for violation of water quality standards—forfeiture of up to $10,000 for each violation for each day in addition to any other judgment and orders of the court. Such a penalty also could be imposed by the court in a case arising from an enforcement conference.

The authority to call an enforcement conference would extend to all water newly covered by the law, to water pollution activity in violation of the standards, and to pollution that endangers the health or welfare of persons. The requirement that a Governor request federal enforcement action in intrastate situations (except shellfish cases) would be deleted. The alleged polluter, as well as the state agencies, would be notified of the remedial action required pursuant to a conference. The second stage in the present three-stage enforcement process, the public hearing, would be eliminated. If remedial action were not taken at the end of the requisite post-conference period (90 or 180 days, as the case may be), or thereafter, the Secretary could request the Attorney General to bring suit to secure performance of remedial action and compliance with the recommendations and applicable standards.

These changes are proposed in recognition of two main facts. There must be a more swift response to water quality requirements than the present enforcement machinery permits; the circumstances of geography cannot be permitted to hamper effective abatement.

Immediate injunctive relief in emergency situations in which there is an imminent and substantial danger to the health or welfare of persons or possible irreparable damage to water quality or the environment could be sought by the Secretary. Authority to require reports, make investigations, subpoena witnesses, require the production of records, enter public or private property for purposes of inspection, and take depositions would be conferred by the bill.

The Administration is committed to a new federalism, to the return of power to the states and the people. The Administration seeks a futher expansion of federal authority in water quality management for overriding reasons—the critical condition of many of the nation's waters, the imminent threat to other waters, the interrelationship of the water resources, and the manifest inadequacy of present law to respond to the environmental crisis.

Conclusion

The federal government viewpoint on the enforcement of water quality requirements differs from, but is not adverse to, that of local government or of the state. Environmental protection is a duty of government at all levels.

The federal government alone has a duty to all of the states and localities, and to all of the American people.

State readiness to meet its primary resonsibilities with respect to the quality of its water assures the state of a greater measure of control over the use and management of its water and the optimum benefits of the Federal Water Pollution Control Act.

Water quality control emerged from early narrow concern with human health to a broader concern with total water resource management, then to concern with water quality protection as critical to the restoration of man's total environment, which the President described in a recent message to Congress as "the most neglected and the most rapidly deteriorating aspect of our national life."

The swift, effective, and equitable enforcement of water quality requirements is a foremost task, if we are "to arrest that decline and begin to revive our habitat."

STATE VIEWPOINT

Thomas C. McMahon

The ability of state regulatory agencies to enforce positively the laws, rules, and regulations of the state is the key ingredient to ultimate success in programs of environmental improvement. The obvious parallel need of financial assistance to communities is recognized, but financial inducements by themselves cannot provide a consistent and equitable water quality improvement program without the ability to obtain adequate criminal and civil remedies in a court of law. This paper will describe statutory powers and administrative regulations of the Massachusetts Division of Water Pollution Control, evaluation of criminal vs. civil proceedings in higher state courts, several pertinent court decisions, and the author's recommendations of means to improve effectiveness in the area of enforcement.

History in Massachusetts

The current Massachusetts program of water pollution control was initiated in September 1966 with the enactment of four major bills into law. One established a $150-mil state construction grant program, including $1 mil/yr for a research and development program designed to develop new and improved waste treatment methods. Another created a new Division of Water Pollution Control under the Water Resources Commission in the Massachusetts Department of Natural Resources with complete authority to implement, administer, and enforce water pollution control within the commonwealth. Two others provided for corporate and local tax incentive assistance to industries installing suitable industrial waste treatment facilities.

Succeeding legislation to date includes an additional $250-mil bond issue for construction grants including prefinancing, a $25-mil low interest loan program for industry, stringent oil pollution laws incorporating emergency clean-up funds, mandatory wastewater treatment plant operator certification, boat pollution and marina control authority, and streamlined administrative enforcement capabilities.

It would seem that the majority of the necessary program components have been established by law, and implementation and enforcement of these laws should give Massachusetts the cleanest water in the nation. Before acceding to this statement, an analysis of the Clean Waters Act, the State Administrative Procedures Act, and the enforcement history of the Division should be made.

Enforcement Provisions of the Clean Waters Act

Chapter 21 of the Massachusetts General Laws includes the enforcement provisions of the Clean Waters Act from Sections 42 through 53. They may be divided into two general categories: criminal citations and civil proceedings against municipal and industrial dischargers. Criminal citations are instituted on request of the Director of the Water Pollution Control Division to natural resources officers who, in turn, investigate and prosecute violation of laws, rules, and regulations in a local district court. Criminal citations normally are requested only in cases of deliberate or willful violation, gross negligence, or extreme recalcitrance to an abatement program.

The civil proceedings are applied primarily to the 130 communities and 400 industries currently on an implementation program adopted by the Water Resources Commission and approved August 8, 1967, by the U. S. Secretary of the Interior. Under the provisions of the State Administrative Procedures Act and Chapter 21 of the General Laws, adjudicatory hearings have been required before the issuance of an administrative order to an industry or municipality. This has been offset by promulgation of consent decrees as an accelerative administrative step. Recent changes now empower the Division to issue orders directly with a 30-day period for appeal to a hearing. In civil proceedings where an administrative order is violated, the Attorney General of the Commonwealth, on request of the Director of the Division, takes appropriate action against the municipal or industrial violator. Action instituted by the Attorney General varies according to the problem but may include injunctions seeking temporary and permanent restraining orders, revised schedules by the Superior or Supreme Court, or significant fines. In cases where court-ordered implementation dates were violated, contempt proceedings have been initiated by the office of the Attorney General at the request of the Director.

Evaluation of Enforcement Actions

Since 1967 there have been a total of 43 enforcement actions taken against municipal or industrial violators of orders or rules and regulations in, or associated with, the Massachusetts Clean Waters Act. Criminal citations have been sought in 21 situations and civil proceedings initiated in 22 incidents. Disregarding for the moment about 10 potential cases that were not admitted to District Court, 75 percent of the criminal cases received reasonable dispatch and resulted in corrective action.

Analysis of the civil proceedings experience of the state is more difficult. Ultimate disposition of most court cases could be considered successful if the time reference point of view is ignored. The Act stipulates "the Supreme Judicial Court and the Superior Court shall have jurisdiction in equity to enforce any orders or determinations of the Divisions." Experience to date with court enforcement shows that the court, rather than enforcing the Division's order, has established new dates for compliance. A typical example of a municipality in noncompliance requiring court orders to proceed reads chronologically as follows:

Town: Blackstone—Blackstone River Basin

	Date	Construction Completion Date
1. Implementation notice sent	December 1967	September 1970
2. Violation—complaint issued	July 1968	
3. Consent decree signed	September 1968	
4. Consent decree violated	December 1968	
5. Referral to Attorney General	April 1969	
6. Marked for trial	August 1970	
7. Disposition	Final Decree, Court Order	March 1973

In this example the administrative procedures required by law for this Division, followed by normal prosecution procedures by the Attorney General, has resulted so far in a 2.5-yr delay. Since September 30, 1970, the town of Blackstone has been in violation of the court order requiring the Division to recommend contempt proceedings by the Attorney General against the town. This, in all probability, will result in either a significant fine or a fourth set of dates. At this time 21 municipalities and industries by virtue of recalcitrance and subsequent litigation are on schedules totaling 43 yr late or an average of 2 yr/violator. Granted, the civil proceedings engender action against the larger and more significant municipal and industrial polluters in the state, it is of importance to recognize that the *Engineering News-Record* construction cost index has been rising recently at a rate of 9 to 10 percent/yr. This unquestionably serves to point out that delays associated with litigation not only are deleterious from a water quality standpoint, but also constitute poor fiscal management at the local or corporate management level and shrink the effectiveness of the state and federal tax dollar used for construction grants.

To criticize the system without recommendation for alternatives is unfair. Nevertheless, the tremendous public demand for environmental improvements deserves a more up-to-date and progressive way of attaining compliance consistent with good management and engineering practices.

Litigation—Help or Hindrance?

The author has stated that the key to environmental improvements is the ability of the regulatory agency to enforce positively the laws, rules, and regulations that have been promulgated. Based on a statistical analysis of court cases, litigation is a time-consuming, awkward, and unpleasant means to an end. To dispel any illusions, it is the policy of the Massachusetts Division of Water Pollution Control to enforce as necessary the laws and regulations on municipal and industrial dischargers where other persuasive and more efficient methods cannot be effected. The abatement and control of wastewater problems is the objective, not filling the courts with lawsuits. Massive litigation is not the answer—*the ability to achieve the goal through litigation, if necessary, is fundamental.* Positive, cooperative methods combined with engineering and construction skill, followed by sound management and operational procedures, provide the best vehicle in water quality management. The adage of bringing the horse to water but not being able to make him drink could be changed to: you can drag them through the courts and make them build a plant, but proper operation and maintenance is best accomplished by mutual cooperation and a sense of environmental responsibility.

A measure of the effectiveness of state water pollution control programs has long been the number of satisfactory court decisions in a state, rather than the number of treatment plants constructed and the overall improvement in water quality. Perhaps a better yardstick would be the largest number of treatment plants built with ensuing water quality improvement for the least time expended in litigation.

Improvements to Enforcement

There recently have been a number of new approaches to environmental improvement in the legal arena. Certain states have initiated a sue now, negotiate later philosophy through actions of the states' Attorneys General. Some have considered the right of the individual citizen to sue, and recently the 1899 Refuse Act has been instituted in certain areas to correct industrial waste discharges. If nothing more, it reflects a growing dissatisfaction at the pace of water pollution control programs in the states and the public desire to institute punitive measures to force compliance. Whether this constitutes a reason for changing the present system is difficult to ascertain. The sue now, negotiate later premise seems to have the basic infirmity of adversely affecting the water quality standards provision of the Federal Water Pollution Control Act as well as state implementation programs. The right of citizens to sue is more complex. If state regulatory agencies, with rules and regulations, laws, sampling and analysis capabilities, and specific legal remedies still cannot effectively persuade the courts to enforce the law, where does the individual citizen stand? The Refuse Act is essentially that and has no place in water pollution control enforcement.

Apparently water pollution control enforcement as an inducive, timely step to satisfactory solutions leaves much to be desired. This is compounded in Massachusetts where strong home-rule laws and sentiment exist, and the determining criteria in towns is successful passage of Town Meeting Articles. For that reason, the time problem is much more severe in dealing with municipalities than with industry in Massachusetts.

Which Way to Go?

The problems previously delineated are expected to be overcome partially by several amendments of the General Laws recently enacted. The fine on criminal citations now has a $1,000-limit instead of $100 as before, and direct orders may be issued without adjudicatory proceedings, but including a 30-day appeal period. This should reduce implementation slippage, but is not likely to overcome fully the other administrative and court delays.

In looking to the future, many treatment plants, interceptors, pumping stations, and outfall sewers will be built without the need for extensive litigation procedures. There will be a substantial number of municipalities and industries, however, that will not provide the necessary treatment facilities without exhausting all administrative and legal remedies dragging out practical environmental solutions for years.

A number of states have created new authorities to accept wastes, and construct, operate, and maintain treatment plants and allied works on a voluntary basis with capital improvements financed by federal and state grants and local contributions—operation and maintenance costs normally coming from user charges based on a volume and strength formula. This unquestionably is a progressive and positive step that virtually would assure improved water quality objectives if the authority is truly a water pollution control entity and is received and utilized by local communities. The theoretical application of such authorities would presumably be more acceptable in the more densely populated states with severe water quality problems.

The issue today is the enforcement of water pollution control. If the "waste acceptance" or "authority" concept is superior from an administrative and water quality improvement standpoint, cannot this be tailored on a mandatory basis for solving the problems of municipalities or industries unwilling to comply under present means?

This method is essentially untried, and unquestionably the mandatory provision will be contested vigorously from local government and in many legislatures.

Good progress is being made, but if 3 yr of enforcement in Massachusetts constitutes a valid pattern, new methods must be sought to improve the system. Water pollution control will only be solved on a broad scale by a continuing, equitable, and timely construction program. If the system has to be changed to accomplish the goal, it should be changed.

LOCAL GOVERNMENT—THE ENFORCER

Allen S. Lavin, Frederick M. Feldman, and Sanford R. Gail

Pollution abatement has now become a politically and socially attractive topic. In reality, the attention is late in coming, and the job is too big for any single agency or level of government. It will take the combined efforts of all levels of government, but especially that of local government.

From the early days of the development of common law, it was the local government that fought against water pollution and attempted to control it. This was only natural because it was the resident of the community who was most vitally concerned with the contamination of the water supply of his town or village. He was the primary victim of the deleterious effluents that were discharged by the local industry into the river or creek from which he obtained his water. Hence, it was logical and understandable that he looked to the governing powers of his community for relief and demanded appropriate action from them.

What was true some hundred years ago is equally true today. The outer manifestations and trappings may have changed but the problem still remains. Though pollution of water has become a thousandfold larger and much more pernicious, the average citizen still looks primarily to the local government for relief. The local governmental agency is the political body with which he is in daily and immediate contact; it is the institution whose workings and deliberations he can observe from close quarters, and it is the agency whose actions he can most readily assess and whose members he elects. State and federal governments are too far removed from the scene to be able to evaluate properly and even understand a local problem that has arisen. Problems of too few personnel and too large a geographical area greatly reduce the effectiveness of state enforcement.

It is important to note that there has been no pre-emption by the federal government in the pollution abatement field. In fact, it is the express intent of Congress that the state and local entities shall have the primary right and responsibilities for the adoption and enforcement of water quality standards.

In a role similar to that of the federal government, state governments and their agencies must act as the overseers and coordinators of activity by the various local agencies scattered throughout the state. A single state pollution control agency is important not only to meet federal requirements, but also to provide uniform minimum standards for the entire state.

Local Role

Local governmental participation, however, is still essential in the effective abatement of pollution. The smaller jurisdictional area allows for a higher saturation of enforcement activities. Local agencies have demonstrated an ability to act more quickly in emergency situations such as the retention and removal of oil spills. Enforcement on the local level is more responsive to the needs of the community and is best able to give the people of the community the quality of water that they desire.

While the state can and should set maximum allowable limits, local communities and their inhabitants have an inherent right to determine the type of environment they want to live in. This right includes the right to adopt and enforce more stringent standards.

Such standards may result in fewer industries locating in a particular community. However, as in many other cases, the prime question is whether the residents are willing to bear the resulting additional tax burden. Therefore, local government plays a vital role in the abatement of pollution and enforcement of the law.

Enforcement Problems

Detection

Detection of violators represents one of the major problems faced by any pollution enforcement agency. To assist agencies in detecting violators, the public must be educated about the visible signs of pollution and encouraged to act as the eyes and ears of the enforcement agency. Detection and investigation are the foundation on which all enforcement is built. The means by which pollution violators are detected must become more and more sophisticated. The Metropolitan Sanitary District of Greater Chicago, in its antipollution campaign, is presently employing the use of helicopters and boats so that large areas can be inspected.

The use of videotape recordings have been employed as evidence in a pollution case. The confrontation of the alleged violator with the visual evidence of his pollution contributed significantly to obtaining a speedy remedy when this matter was brought before the court.

As new detection techniques are developed, the District has adopted them as quickly as possible. A study of the use of infrared photography to locate pollution coming from submerged outfalls is under way, including the possibility of using radioactive and chemical techniques to detect and trace the sources of pollution.

In addition to the problems relating to detection, repeated experiences have shown laboratory standards to be obsolete.

The District, as most other enforcement agencies, uses as its handbook for performance of sampling and chemical analyses "Standard Methods."[1] However, experience has shown that "Standard Methods" is not very specific with regard to proper sampling techniques. Furthermore, it should be revised more frequently as new, reliable chemical detecting and testing equipment is developed and becomes commercially available. For example, in the analysis of metals, "Standard Methods" has not yet adopted the atomic absorption spectrophotometer, which is easy to use and gives reliable results quickly.

Standards

Another significant problem is that of standards. Stream criteria are nearly impossible to enforce because it is extremely difficult to show a direct relationship between the effluent of one company and the quality of the water in the stream. This is especially true when dealing with already polluted water. Stream criteria are best suited for use as goals or the optimum levels to be achieved in cleaning up polluted water but are too flexible and do not lend themselves to strict interpretation.

Effluent criteria, on the other hand, are a better means of effecting compliance by individual polluters. Such criteria stated in terms of milligrams per liter lend themselves to use as an enforcement tool because of the fact that a company's discharge of pollutants can be reduced to numerical values and readily compared with established effluent standards, but even this creates problems.

A standard does not account for the volume of effluent discharged by the polluter. Milligrams per liter alone can be unfair where there are two polluters discharging the same pollutant but in different volumes of water.

In fact, one can circumvent this type of standard by just adding more water. Fifty years ago dilution was the primary means of treating wastewater and other pollution, but today thinking has changed although concentration of a pollutant should not be discarded totally as a criterion.

In order to protect the quality of water from large volume discharges better, new standards must be established that would specify the limits of the discharge of pollutants in not only milligrams per liter but also milligrams per liter not to exceed a fixed volume by weight.

The profession must develop standards for the abatement of pollution that more clearly reflect the problem and are more just in their application. A standard based on milligrams per liter not to exceed a specified volume per gallon or per day is the right approach. But the legal profession needs assistance in developing these standards and adopting them for use in the protection of water resources.

Judicial Remedies

Although there are many statements to the contrary, the proper forum for compelling pollution abatement is the court, and although some polluters work diligently at the administrative level, it is found most often that abatement action increases and compliance is achieved only after a suit if filed and the polluter is summoned before the bar of justice. It is interesting to note that court action tends to make subsequent administrative procedures more effective.

Members of the judiciary have become keenly aware of the problems of pollution. In the Chicago area judges are very vocal in their efforts to assist agencies in compelling alleged polluters to comply with the applicable standards. The hard line that some judges have taken and the publicity that accompanies a judicial proceeding have worked to the advantage of enforcement agencies. Many polluters, knowing the level of prosecution success in court, have tried harder and have successfully achieved compliance at the administrative level in order to forego the possibility of judicial proceedings and the attendant publicity.

Injunctions

In spite of the increasing awareness of the judiciary, the enforcer still encounters philosophical and technical problems in the prosecution of pollution suits.

In this respect, there should be a realigning of the burden of proof because the polluter is in the best position to know its production facilities and the content of its effluent. Therefore, it should be industry's burden to show that its effluent does not pollute.

Paralleling this are the problems that arise when courts invoke the equitable principle of "balancing the equities," which simply means equating the relief to be obtained, if an injunction is issued, against the hardships that may be incurred. Often an injunction against a discharge of pollutants is equivalent to an order to shut down the facilities. If this is required to obtain compliance, then this is what must be sought. Courts, however, seem inclined to give more weight to the economic hardship to be sustained by the polluter and seem to discount the adverse effects of the polluter's action on the water. Many judges find it repugnant to shut down production for the purpose of stopping water pollution. Yet, industries can be shut down by a labor strike. Management has, itself, shut down production and locked out its employees solely over questions of money. But when a manufacturing plant is threatened with a shutdown because it is polluting the water it uses and discharges, a cry goes up that people are being deprived of their livelihood. The adversity of shutting down a plant that pollutes the water is much less onerous than an industry-wide strike when one considers the necessity for and benefits of cleaner water. A change in thinking and philosophy is necessary on the part of the courts. Although they know of the problems that pollution presents, it is yet to be shown that the courts truly realize its disastrous consequences.

Perhaps in weighing the equities, the balance should be tipped toward the presumption of harm to the waterways to offset the polluter's unique knowledge of what he discharges. Furthermore, the judicial perspective must be changed to consider the cumulative affect of all pollutants present in the same body of water. The Chicago experience has shown that courts will act strongly in pollution matters where immediate danger to human life is shown. The greatest task faced today is to convince the courts that harm to the water is as significant a problem as direct immediate injury to a person.

Receivership

Many pollution violations occur when industrial production outstrips the design capacity of existing pollution control facilities. In cases such as this, the courts should order that production be curtailed so that the existing treatment plant can cope with the waste until such time as treatment facilities are enlarged. Further, where a polluter is operating at a profit and the court is hesitant to close it down, the corporation should be placed in receivership and the profits of production applied to pollution abatement until the polluter is in compliance. *It is unethical and inequitable to pollute at a profit.* Under this receivership concept the receiver would oversee the company's pollution abatement activities. Under court supervision the receiver would insure that corporate funds are being applied to the abatement efforts. The receiver would direct the company's activities so that pollution control programs get as much attention as production. It is very difficult to solve an around-the-clock pollution problem during an 8-hr shift. If production continues 24 hr/day, then pollution abatement activities, too, must continue on a 24-hr/day basis.

Fines

Another deterrent that should be readily available to an enforcement agency is the levy of fines against proven polluters. Only recently has new legislation been passed establishing fines that make it unprofitable to pollute. Reason dictates that, except for the adverse publicity, token or nominal fines such as $100/day are often more willingly incurred than the spending of substantial sums for the permanent abatement of pollution.

To reverse the trend, a presumption should be established that once a polluter has been proven to be polluting, it should be presumed that such pollution continues from that day until such time as the polluter can satisfy the court that the pollution has been abated and the violator is now in compliance.

Jailing

A final deterrent to pollution is to jail the polluters. Few people, if any, equate a polluter with an arsonist or murderer. An arsonist can be imprisoned for life and a murderer can be executed. Certainly one who discharges pollutants into the water of this nation, thereby reducing the utility of the water and its ability to cleanse and regenerate itself, commits a crime. The party that places pollutants in the nation's water, thereby reducing the available water supply, is as much a criminal as a man who gives adulterated candy to children.

Many companies amass vast fortunes and are willing to pay nominal fines in lieu of abating pollution. Nevertheless, jail a corporate officer who allows or is in a position to know that his firm is allowing pollutants to reach the water, and he will quickly come to realize how important clean water is and how his company can best eliminate its problems.

The ever-present threat of jail for a polluter and its corporate officers is a stern measure but one that is needed to preserve the birthright of clean water. Hopefully, this threat will cause industries and their officers to realize that everyone needs clean water and that they have a great deal to say as to how much time people will spend on this earth.

Conclusion

This country was founded on the principle that all men are endowed with certain inherent and inalienable rights; among them are life, liberty, and the pursuit of happiness.

Implicit in these rights is the right to have clean air to breathe and clean water to drink and use. This society was conceived to protect those rights. Today, society must renew its dedication to preserving the rights of every individual to have and enjoy clean air and water.

Just as Lincoln spoke of the Civil War in his Gettysburg Address, we, too, are now engaged in a great war against pollution, a war that is testing whether this nation or any nation so conceived and so dedicated can long endure.

The fight against pollution is a war! It is a war we must not lose!

Reference

1. "Standard Methods for the Examination of Water and Wastewater." 12th Ed., Amer. Pub. Health Assn., New York, N. Y. (1965).

INDUSTRY'S VIEWPOINT

H. Edward Dunkelberger, Jr.*

There is no single or uniform industry viewpoint regarding federal water pollution control enforcement. There has been, however, an increasingly widespread concern in the business community about somewhat unorthodox enforcement approaches to federal water pollution control, which have little or no relationship to the carefully constructed enforcement procedures established by Congress in the 1965 amendments to the Federal Water Pollution Control Act, and are described by Stein in his paper.

The Certification Problem

The first of these two new theories of federal enforcement is an outgrowth of amendments to the Federal Water Pollution Control Act enacted by Congress in early 1970.

These amendments added a new section to provide that any applicant for a federal license or permit to conduct any activity that may result in any discharge into navigable waters must first provide the licensing agency with a certificate from the state in which the discharge originates. That certificate must indicate the state water pollution control agency's belief that there is reasonable assurance that the activity will be conducted in a manner that will not violate applicable water quality standards.

Although at first glance this provision may sound quite innocent—merely an advance guarantee by industry that its facilities will not run afoul of minimum stream standards—many observers believe that the certification requirements will place a formidable new obstacle in the path of future development of much needed manufacturing, wastewater treatment, and electric power generating facilities.

For plants currently in operation and subject to the provisions of the Act, the burden is on the government to take court action against an alleged violation of applicable water quality standards in accordance with detailed procedures set forth in the Act. Until the 1970 amendments, it was the responsibility of a company or municipality planning a new facility to decide for itself what wastewater treatment procedures would be necessary to achieve compliance with water quality standards.

* The author was assisted by Peter M. Phillips of Covington & Burling in the preparation of this paper.

With the enactment of the 1970 amendments, the builder of a new facility now must first obtain certification from the state or interstate administrative agency charged with the local enforcement of water quality standards. Thus, prior to commencing construction, the owner must affirmatively provide reasonable assurance that no violation will occur during the plant's future operations.

The Act directs the states to establish certification procedures, including provisions for public notice of all applications and for public hearings, when deemed appropriate by the state agency. Consequently, the plant owner may be forced to endure significant delays while the technical question of whether proposed treatment procedures will or will not result in compliance with stream standards is debated in the public forum.

Although there is no basis for believing that state authorities will be unreasonable in providing certification, it is clear that any good faith difference of opinion as to the adequacy of proposed treatment facilities will be resolved in favor of the state, which is in position to refuse certification and completely stymie the planned construction. The licensing federal agency would be powerless to grant the license if the state denied the requested certification.

The only safeguard written into the Act is that if the state refuses to act on a requested certification within a reasonable time—no longer than a year—the certification requirement is waived. There is no statutory provision for judicial review, although the legislative history suggests that Congress believed that court relief could be obtained when a state had acted arbitrarily or capriciously in refusing certification.

This certification procedure applies whenever the builder of a new plant or other facility must obtain a license or permit for either its construction or operation. Obviously any power-generating plant licensed by the Atomic Energy Commission (AEC) or the Federal Power Commission will be covered. In addition, virtually all new construction on navigable waters will be covered because of the need to obtain a permit for such construction from the Army Corps of Engineers under the 1899 Rivers and Harbors Act.

Applicable Standards

A number of questions will undoubtedly arise under this new requirement that federal licensees obtain certification from the appropriate state that a proposed facility will comply with "applicable water quality standards." One of these questions almost certainly will involve the meaning of the term "applicable water quality standards."

There should be little difficulty in this regard if the state has adopted water quality standards for the body of water in question and the federal authorities have approved them under procedures of the Act. Problems may arise, however, if the state has adopted standards that have been disapproved or not yet approved by the federal government, and federal standards have not been promulgated in their place.

Imagine the plight of the company or municipality that is preparing to construct a new facility on an interstate stream for which the state has adopted standards that were disapproved or not yet approved by the federal government. If the unapproved state standards are not deemed to be the "applicable water quality standards," then under the Act the builder need not obtain certification, but once the licensee is notified of the subsequent adoption of "applicable water quality standards" he must comply within 6 months or face suspension of his license.

Under this interpretation a well-intentioned company might begin construction of a facility meeting all known water pollution control requirements, including newly adopted state standards, and then be forced to suspend operations or refit its plant at substantial cost and delay as a result of the adoption of new federal standards for which it had neither sufficient advance warning nor opportunity to prepare. This hardly would be a just result.

Fortunately, there is substantial evidence in the language of the Act and in the legislative history of the 1970 amendments to the effect that unapproved or rejected state water quality standards are "applicable water quality standards" nonetheless.

For example, it seems clear that state standards for *intra*state waters are deemed to be "applicable water quality standards" for purposes of certification, even though the federal government is authorized to approve and enforce only standards for *inter*state waters. It is logical to conclude then that unapproved, but otherwise lawful and effective state standards for *inter*state waters, are the "applicable water quality standards" for certification purposes, if at the time of certification federal standards have not been adopted for those waters.

This evidence that state standards rejected by the government are in fact the "applicable water quality standards" for purposes of certification militates strongly against requiring industry to conform to standards approved "after the fact" in those cases where state water quality standards are actually in existence—though not federally approved—at the time certification is requested.

Operating Permits

Facilities requiring operating permits from federal agencies following completion of construction, such as those involving the use of nuclear power, face an additional hurdle. Although certification may have been received from state authorities, and a construction permit granted by the applicable federal agency at the time the project was commenced, an operating license from the AEC may be denied if the certifying authority notifies the AEC:

> that there is no longer reasonable assurance that there will be compliance with applicable water quality standards because of changes since the construction license or permit certification was issued in (A) the construction or operation of the facility, (B) the characteristics of the waters into which such discharge is made, or (C) the water quality standards applicable to such waters.

While changes in the facility or the water may provide a reasonable basis for a reevaluation of certification, the third condition, changes in the standards, places companies in the impossible position of trying to anticipate changes in water quality standards during the course of construction or face denial of an operating license for a plant built in conformity with all requirements existing at the time certification was granted originally. This harsh result could well bring about significant financial reverses for well-intentioned companies and municipalities. It seems clear that consistency and continuity of certification and licensing requirements by both state and federal authorities must be maintained to prevent imposition of unjustifiable requirements on owners of new facilities.

The *Qui Tam* Problem

Currently enjoying a degree of attention not received before in its 71-yr history is the Refuse Act of 1899, the basis for the second new theory of federal water pollution control enforcement referred to in the opening paragraph. The Refuse Act is a companion statute to the 1899 Rivers and Harbors Act.

Section 13 of the Refuse Act prohibits the discharge of matter of any kind into the navigable waters of the U. S. unless flowing from streets or sewers in a liquid state. For most of its history this Act has been used solely to prosecute those responsible for discharges that adversely affect navigation. But in recent years there have been increasing efforts to use the very broad language of this Act to attack all such discharges that may have an adverse effect on water quality.

By its terms, the Refuse Act prohibits the deposit of refuse matter in any navigable water of the U. S. unless a permit authorizing such dumping has been obtained from the Secretary of the Army. Persons violating the Act are criminally liable to a fine of $500 to $2,500 or imprisonment for from 30 days to 1 yr, or both. The court is given discretion to direct payment of one-half the fine to the person furnishing information leading to conviction.

In addition, it may be possible for the government to prosecute a civil action under the Act to prevent further prohibited discharges.

However, if used for non-navigational purposes, i.e., for chronic pollution abatement, it is inevitable that actions under the Refuse Act will undermine and perhaps directly conflict with pollution abatement procedures under the Federal Water Pollution Control Act.

Realizing this, the Justice Department, in July 1970, issued "Guidelines for Litigation under the Refuse Act" to the several U. S. Attorneys. The Guidelines state that:

> The policy of the Department of Justice with respect to the enforcement of the Refuse Act for purposes other than the protection of the navigable capacity of our national waters, is not to attempt to use it as a pollution abatement statute in competition with the Federal Water Pollution Control Act.

Rather, the U. S. Attorneys have been directed to use the Refuse Act to combat "significant discharges that are either accidental or infrequent, but that are not of a continuing nature resulting from the ordinary operations of a manufacturing plant." Discharges of a continuing nature are viewed as the type "that the Congress created the Federal Water Quality Administration to decrease or eliminate."

Although government use of the Refuse Act as a pollution abatement device thus has been put in its proper perspective, the Act may pose an additional problem.

Several cases have now been brought by interested citizens who claim that the 1899 Act authorizes them to bring private actions against polluters. However, the use of the Refuse Act for such private actions, termed *qui tam* actions, never has been upheld authoritatively and is open to question.

Proponents of the *qui tam* actions under the Refuse Act acknowledge that "the Act does not explicitly state that citizens have such a right." However, they counter by stating that "it does not explicitly deny to citizens this right," and cite cases allegedly supporting the proposition that "where a statute providing for a reward to informers does not specifically either authorize or forbid the informer to institute a *qui tam* action, such statute is to be construed as authorizing such suit."

Investigation has shown that, under American law, the courts generally have held that the *qui tam* action may be brought by an informer only under a civil statute that affirmatively authorizes him to sue to recover a share of the statutory penalty or forfeiture to which he is entitled. Furthermore, it is an established legal principle that the burden of showing that a statute has conferred the right to bring a *qui tam* action rests on the informer. No such authorization appears in the Refuse Act or its legislative history.

The Refuse Act is an exclusively criminal statute that speaks in terms of government prosecution leading to the imposition of a fine, imprisonment, or both. Research has failed to disclose a single American case in which a *qui tam* action has been asserted successfully to enforce a criminal statute by collecting a criminal fine.

Rather, *qui tam* actions have been used extensively to recover civil penalties or forfeitures, but the Refuse Act makes no provision for the maintenance of either a private or a government civil action to collect a penalty or forfeiture. Thus, any *qui tam* action under the Refuse Act of 1899 is open to serious legal question.

Apparently many Congressmen have their doubts about the adequacy and legality of the Refuse Act *qui tam* approach because numerous bills designed to establish private rights of action to abate pollution were introduced in both houses of Congress in 1970.

Conclusion

The author does not challenge the government's right or responsibility to get on with the very necessary job of abating water pollution and does not claim that industry is totally blameless for the current condition of many lakes, rivers, and streams. However, the viewpoint of industry seems to be that federal enforcement of water pollution control requirements must be reasonable and consistent and based on the comprehensive statutory scheme set forth in the Federal Water Pollution Control Act. Piecemeal private or government enforcement based on novel principles of questionable legality cannot hope to achieve a meaningful degree of water pollution control.

Wastewater Treatment Plant and Sewer Construction Cost Index, October 1970 (1957–59 = 100) Federal Water Quality Administration

City No.	Cities	WPC–STP Values	Change Since October 1969, (%)	WPC–S Values	Change Since October 1969, (%)
01	Atlanta	135.79	+11.1	136.69	+12.4
02	Baltimore	139.20	+10.1	145.86	+13.9
03	Birmingham	121.29	+ 6.2	123.01	+ 4.7
04	Boston	150.44	+ 7.1	156.18	+ 5.2
05	Chicago	153.99	+12.3	153.95	+13.0
06	Cincinnati	147.65	+ 8.6	160.11	+ 9.1
07	Cleveland	155.43	+ 6.7	165.67	+ 7.4
08	Dallas	130.37	+ 6.1	123.87	+ 5.3
09	Denver	134.68	+ 7.0	131.45	+ 1.7
10	Detroit	162.53	+ 6.6	171.91	+15.4
11	Kansas City	154.45	+21.8	164.05	+21.4
12	Los Angeles	148.77	+ 8.2	157.37	+10.1
13	Minneapolis	153.86	+11.8	163.36	+12.2
14	New Orleans	138.00	+10.5	140.71	+ 5.7
15	New York	168.72	+ 8.8	175.89	+ 7.8
16	Philadelphia	148.07	+ 9.5	155.22	+ 4.3
17	Pittsburgh	154.25	+ 5.0	161.17	+ 5.8
18	St. Louis	153.11	+ 9.1	159.76	+11.2
19	San Francisco	157.29	+ 6.9	174.08	+10.2
20	Seattle	153.27	+ 6.3	167.38	+ 5.6
	National Index	148.07	+ 9.0	154.38	+ 9.1

STP = Treatment plant index.
S = Sewer index.

Chlorination of water as carried out by the Metropolitan Water Board

by A. H. Cooke

The daily quantity of water supplied now approaches 400mgal. ($1{\cdot}84 \times 10^6$ m^3) of which 68 per cent is derived from the River Thames, 17 per cent from the River Lee, and the remainder from wells. Both river sources are polluted by land washings, treated sewage and trade wastes, but the well waters are mostly of the highest bacteriological and chemical quality.

The river waters are stored for weeks or months in 39 impounding reservoirs, with a capacity altogether of 30,000 mgal. (136×10^6 m^3), during which time much suspended matter settles out, most pathogenic bacteria and many undesirable organisms die and variations in quality are evened out. On the debit side, algal growths can produce filtration and taste problems. The water then gravitates to 11 filtration stations of which the largest is Hampton in the Thames Valley, pumping 120mgal./d (552,000 m^3/d).

The new Coppermills Works (Lee Valley) will, when completed, pump 108mgal./d (498,000 m^3/d).

At the filtration stations, the water passes through rapid gravity sand filters (primary filters) at the rate of 20 ft/h (6m/h), and then through slow sand filters (secondary filters) at a rate of 8in/h (20cm/h). The primary filters are backwashed daily by an upward air scour followed by water, and the secondary filters are cleaned when required by draining down and skimming off 0·75in (2cm) of top sand.

At Ashford Common (Thames Valley), pumping 90mgal./d (334,000 m^3/d) and at Lee Bridge (Lee Valley), pumping 45mgal./d (217,000 m^3/d), stainless steel micromesh strainers, capacity 4mgal./d (18,400 m^3/d) each, are used in place of rapid gravity filters. Both serve to relieve the load in the secondary beds, but the rapid gravity sand filters, in addition, exert a marked biological effect, particularly by oxidising ammonia nitrogen to nitrate nitrogen.

Slow sand filters quickly develop a surface skin, consisting of silt, algae and organic debris which gives a very fine filtering action capable of removing bacteria; the filter skin and the sand beneath also exert a biological action. After filter beds have been skimmed and returned to service, it is usual for a day or so to elapse before the best filtering quality is achieved. During cold spells, when the temperature is less than 4°C, the oxidising bacteria in the filter beds are less active, and some ammonia can come through with the filtrate. Diurnal changes on hot summer days can produce somewhat similar effects, and chlorination problems arise in both situations (see later).

After final filtration, the water is pumped into large concrete contact and balancing tanks, in which the disinfecting chlorine is added, and the water retained for 1h or more. Waters from the 32 well stations used for direct supply are, with two exceptions, pumped directly from wells or boreholes into contact tanks without prior filtration, the retention time in these cases is 20min or more.

General principles of chlorination

Like all chemical disinfectants chlorine obeys the two main laws of disinfection, namely, the larger the dose the greater the kill and, the longer the time of contact with the bacteria the greater the kill; these criteria govern the method of chlorination.

It is necessary to have a full appreciation of the forms chlorine takes in water. If present in the free state, for example, as in hypochlorous acid, it is a very powerful germicide, killing vegetative bacteria in a matter of seconds.

If, however, ammonia or similar compounds are present, they combine with the chlorine to form chloramines and the residual chlorine measurable in the water is often termed 'combined chlorine'. Chloramines are much slower in their disinfecting action; only achieving in hours what free chlorine can do in seconds. They are, however, stable compounds so the slight bactericidal property persists for a considerable time.

Where the water is obtained from rivers containing treated sewage effluents and industrial wastes, almost all the undesirable bacteria are removed by storage in reservoirs and by sand filtration, so that few remain by the time chlorine is added. Chlorination of the Board's supplies is terminal and is therefore a final safeguard or a 'finishing off' process. Chlorine is introduced at the inlet to the contact tank in sufficient amounts to provide a predetermined residue of free chlorine after a prescribed period of time. The standard residue for this particular period of time has to be determined for each works and depends, among other factors, upon the layout of the filtration station, rate of flow of water and the bacteriological and chemical quality of the filtered water. It is known by experience that with this arrangement the remaining undesirable bacteria are destroyed by the chlorine in a few minutes. As water passes through the contact tank, chlorine continues to be consumed by traces of organic matter still present in the water, so that the chlorine is usually hardly detectable at the end of the contact period. This is self induced dechlorination and a safe, clear and bright, tasteless and odourless water has been produced.

When the temperature of the water falls below 4°c the nitrification processes break down at sewage works, in the rivers and in the filters so that the filtered water contains increasing amounts of ammonia. The removal of this ammonia by the oxidative action of chlorine is often impracticable on account of the high dosage required and the large scale plant that would have to be installed for only short periods of days one or twice a year. At such times the method of chemical treatment is to utilise the natural ammonia by adding sufficient chlorine to produce adequate concentrations of chloramine which will effectively treat the water during the contact period of 1h that is provided. The concentration of chloramine produced may be too high to pass into supply as the taste and odour would lead to complaints from consumers. Therefore sulphur dioxide is provided at the end of the contact tank to reduce the concentration of residual chloramine to a level tolerable to the public.

It is often desirable for the water pumped into supply to contain a trace of chlorine in a form not perceptible by taste to the consumer, *i.e.* as monochloramine. When no natural ammonia is present this is brought about by the addition of a very small dose of ammonia to the water towards the end of the contact period; it may also be necessary to add a little more chlorine at the same time.

It is vital that chlorination should be continuous and accurate, and so adequate standby plant must be provided and special precautions taken to avoid any interruption in the chlorination process.

Chlorine dose required for disinfection

Research by Griffin,[1] Fair,[2] Taras,[3] Palin[4] and others has greatly improved our knowledge of the processes of disinfection by chlorine. With filtered river water organic matter is always present but, except for cold spells and a continuous period of summer days, simple ammonia nitrogen is oxidised and is not a problem. When the chlorine is added, some 40 per cent is, under normal conditions, deviated in the first 10min by easily oxidised matter in the water and a slower rate of reaction continues until the water reaches the end of the contact tank. A typical example would be that a dose of 1·3 p.p.m. gave 0·8 p.p.m. 'free' chlorine after 10min and 0·2 p.p.m. 'free' chlorine + 0·1 p.p.m. chloramines after 2h.

In very hot weather, this situation can be greatly changed by diurnal variations in the water, thought to be brought about as follows:—hot sun on the secondary filter beds causes vegetation to produce a great deal of nascent oxygen, so that the water is supersaturated with oxygen which appears to take the place of the chlorine normally lost in the first 10min. But this does not affect the subsequent slow organic demand, *i.e.* to obtain a free chlorine residue of 0·8 p.p.m. after 10min may only need the addition of a 0·9 p.p.m. dose during the afternoon. When the sun goes down, oxygen production ceases and anaerobic reduction sets in and by the middle of the night, allowing for a 6h delay for the water to pass through the beds, the 10min chlorine demand may increase to 2·0 p.p.m., *i.e.* to obtain the '0·8' after 10min may require a dose of 2·8 p.p.m., which will still be followed by a slow loss in the contact tank. By 6.0 a.m. ammonia may also appear and further complicate the issue by forming monochloramines and so on. The position usually reverts to 'normal' by midday. The chlorinator operators know what must be done when this happens, and the mechanical cleaning of filter beds is now reducing the incidence of this effect by speeding up the cleaning process.

The following table gives some representative figures for the Board's river and well waters which are of use in considering chlorination levels:

Table

	pH	*Ammonia nitrogen*	*Albuminoid nitrogen*	*Colour 24in tube*	*Turbidity units*
River waters	8·1	0·18	0·29	37	21
Filtered river waters	7·9	0·02	0·09	14	0·1
Chalk wells	7·2	0·01	0·02	1	nil

From the point of view of disinfection and taste, the two important differences between ammoniacal nitrogen and albuminoid nitrogen arise from their combining values with chlorine and their times of reaction. In the case of simple ammonium salts, the reactions which take place with chlorine are well known and can be mathematically predicted, but with organic nitrogen the reactions are less well defined and proceed slowly.

Reactions between chlorine and simple ammonium compounds

Chlorine dissolves in water to form hypochlorous acid:

$Cl_2 + H_2O \longrightarrow HOCl + HCl$

If ammonia is present, monochloramine is formed within seconds when the pH exceeds 7·0:

$HOCl + NH_3 \longrightarrow NH_2Cl + H_2O$

Monochloramine is very stable, tasteless and a slow disinfectant compared with free chlorine. The combining ratio for the reaction is 5:1 chlorine to nitrogen.

Should more chlorine be available, a further reaction takes place to give dichloramine at a pH of over 7·0.

$NH_2Cl + HOCl \longrightarrow NHCl_2 + H_2O$

This second reaction takes place more slowly, taking some 20min or more for complete reaction in the pH range between 7 and 8. Dichloramine has an unpleasant chlorinous taste and, like monochloramine, is a poor disinfectant. When the ratio of chlorine to ammonia nitrogen is about 10:1, a point of minimum total residual chlorine, known as the 'break point', is obtained, provided sufficient time is allowed for the reactions between chloramine and free chlorine to take place. When the ratio of chlorine to ammonia nitrogen exceeds 10:1, small quantities of nitrogen trichloride are formed by the reaction of the extra chlorine with dichloramine, *i.e.*

$$NH_2Cl + 2HOCl \longrightarrow NCl_3 + 2H_2O$$

The Board's Hadley Road well, which has a naturally occurring ammonia nitrogen content of 0·2 p.p.m., and a pH of 7·4, lends itself to breakpoint treatment. Dosing at 2·5 p.p.m. produced a free chlorine residue of 0·5 p.p.m. at the end of the contact tank, with traces only of chloramines; no ammonia nitrogen remained.

Dechlorination by ammonia

Dechlorination of water can be induced by adding ammonia in the correct concentration to achieve 'breakpoint' conditions, but if organic matter is present control may be difficult. A more usual procedure is to add ammonia to convert all the residual chlorine to monochloramine.

Chlorine and organic compounds

In contrast with the predictable behaviour of chlorine with simple ammonia nitrogen, organic nitrogen reacts in a manner dependent on the nature of the substances present and generally the process is slow. Compounds similar to mono- and di-chloramines are formed and traces of nitrogen trichloride.

The main purpose of disinfection must be first and foremost to kill off all pathogenic bacteria and other undesirable organisms. Coupled with this is a duty to the public to produce a palatable water, for traces of decayed vegetation and, in particular, simple or complex phenolic substances produce unpleasant tastes in water when it is chlorinated. As little as 1 part of phenol in 10^9 parts of water produces a medicinal flavour when marginally chlorinated, but fortunately an excess of chlorine usually clears up the taste or transforms it to a more acceptable, and removable, chlorinous taste. All chlorine compounds are dechlorinated by sulphur dioxide and chlorinous tastes disappear at the same time (9 parts of SO_2 remove 10 parts of Cl_2). It can be theoretically argued that the best level of chlorination of a water containing organic matter is just greater than the total nitrogen breakpoint so that the minimum quantity of undesirable dichloramine and nitrogen trichloride are formed.

Chlorination at the Metropolitan Water Board's works

The early years of chlorination at the Metropolitan Water Board are described by Byles.[5] In more recent times, prior to the 1939-45 war, disinfection was by chloramination, ammonia being added as ammonia gas or ammonium sulphate solution before the addition of chlorine. There were no contact tanks and reliance was placed on contact in the mains. During the war, chlorination of the water prior to filtration (pre-chlorination) was introduced as a safety precaution, but it produced problems and complications so that during and immediately after the war the treatment was changed to free residual chlorination at the filtration plants, where contact tanks were constructed to hold up the water for at least one hour in order to achieve adequate disinfection before the water was pumped into supply. Pre-chlorination was also abandoned in order to permit the sand filters to perform their natural processes. Chlorine is most effective after final filtration where it produces less complications and less is needed to give the maximum effect. In the case of the well stations the method of superchlorination followed by dechlorination with sulphur dioxide after contact for 20min was introduced.

River derived supplies

As the water enters the contact tank the strong chlorine solution, of the order of 1000 p.p.m., produced by the chlorinators is injected into the main stream of water on its way from the filters. The chlorinators are supplied with gas from banks of drums each containing 17cwt (865kg) of liquid chlorine. Two or more chlorinators share the treatment supplied from two separate header lines as a precaution against failure in one instrument. The drum room and instrument room must be heated, in the case of the former to overcome the effect of cooling as the liquid chlorine expands to a gas, and in the latter to prevent reliquefaction in the gas lines which causes impurities in the chlorine, such as hexachlorethane, which deposit and block orifices. The gas lines are made of iron and the solution lines of plastic material.

After a short period of contact, usually 5 to 10min, facilities are available for a sample of the treated water to be pumped back to the instrument room for chemical testing and automatic recording and control purposes. At each station there is a specified free chlorine concentration required after a given period of contact. The dosing is adjusted to give the required level of residual free chlorine, as determined by chemical test. After the specified contact the test sample flows through a residual recorder which is adjusted to give the correct reading, and thereafter any changes in the residual free chlorine value which take place, due to pumping rates or quality variation, are automatically compensated by the residual chlorine controller. On receipt of signals from the recorder, the controller, by vacuum or electrical adjustment, changes the gas flow from the automatic instruments. This change is made incrementally and will normally control the dosing to keep the residual value within $\pm$ 0·03 p.p.m. of that prescribed. Alarm signals are given if the recorder fails to keep the residual level steady and the attendants make manual adjustments. To cover extreme requirements the chlorinators can add up to 5 p.p.m. of chlorine at full pumping rate.

When the water reaches the end of the tank, it is re-examined and if necessary the residual chlorine level is adjusted. It may be lowered by adding sulphur dioxide, or increased by adding more chlorine, or converted to chloramine by adding ammonia gas with or without chlorine, before pumping into supply. This terminal treatment is usually applied manually, but can also be residually controlled, as at the Board's Middlefield Road Well where sulphonation is controlled by an automatic residual controller.

Drums of sulphur dioxide are usually separately housed but could be kept together with the chlorine drums. On the other hand ammonia cylinders must be kept separate from the chlorine in order to avoid any chance of the production of explosive nitrogen trichloride in case of leaks. All chlorine lines and drums are painted yellow and sulphur dioxide lines and drums are painted green. Ammonia cylinders are black with red tops.

It is easier to look after the disinfectant dose at the start of the tank and to adjust it at the end than to try to do the whole thing in one operation.

Treatment of well waters

Well waters are usually treated with a fixed dose of chlorine, either 0·5 or 1·0 p.p.m., and completely dechlorinated at the end of the contact tank after 20 to 30min. Both chlorine and sulphur dioxide are flow controlled by means of a venturi, the sulphur dioxide dose being balanced against the chlorine dose. Chlorination of the wells is in most cases precautionary treatment against sudden pollution. A number of northern area wells which have an excellent bacteriological record and contain native ammonia, are chlorinated and the resulting chloramine pumped into supply without dechlorination after contact.

With filtered river waters, experience indicates that total residual chlorine of 0·3 p.p.m. leaving the works will not produce complaints, but above this level many people object to the flavour. In the case of the Kent wells, as little as 0·05 p.p.m. of free chlorine left in the supply water can result in individuals complaining. It seems likely that the 'chlorinous' tastes are produced in the reticulation system.

The bacteriological quality of most well waters is so good that the greatest care has to be taken not to produce water after chlorination and dechlorination which is worse, in terms of bacterial content, than the raw water, due to ingress of airborne sporing organisms or bacteria in pump packing, and so on.

The annual consumption of chlorine is 775 tonne and sulphur dioxide 80 tonne. There are more than 100 separate points in the undertaking at which chlorine, sulphur dioxide or ammonia are administered and have to be maintained.

Control of chlorination

The essential feature of successful chlorination is that water free from any undesirable bacteria, that is to say of satisfactory hygienic quality, is produced. At the same time every endeavour should be made to avoid any detectable taste or odour in the water. Experience shows that the Londoner is generally very sensitive to chlorinous tastes and odours and successful chlorination is often difficult to achieve. As already stated, the vital essential of chlorination is that it must be continuous and accurate. Therefore at all stations there is adequate standby plant and at filtration stations chlorination attendants work on a shift system so that the plant is watched continuously day and night. As well as watching and maintaining the instruments, they check the automatic recording equipment by making chemical tests for residual chlorine at frequent intervals from selected points in the mains and contact tank. They keep careful records of the amount of chlorine, ammonia and sulphur dioxide used, the pumping rates and the residual chlorine estimations. Constant control of dosage of chlorine is dealt with centrally from the main laboratories and under the general direction and responsibility of the director of water examination.

In contrast, a number of well stations are unattended at night and have only a day man; others are only visited occasionally. At such places fail safe arrangements are in operation, the residual recorders being set to cut out the pumps if the residual chlorine drops below a predetermined level, and before this point warning signals are transmitted to the parent attended stations. When the dosing is 1·0 p.p.m., cut-outs are set at 0·65 p.p.m. and warnings at 0·8 p.p.m. Where the dosing is 0·5, cut-out is at 0·25 p.p.m. and warnings alarm at 0·4 p.p.m., no chlorine being in the normal way deviated by these well waters.

Chlorination supervisors with radiotelephone communication visit all stations daily and are responsible for minor repairs to apparatus, adjustment of recorders, installation of new equipment, technical supervision of the attendants, and so on. In the case of a well cut-out, if it is due to a fault in the chlorination, the standby supervisor for the area rectifies the fault and restarts the pumps. To do so a time delay arrangement has to be switched in to overcome the cut-out until freshly chlorinated water reaches the cell, usually in less than 10min. Fortunately cut-outs are usually due to power failures and so on, and not to an increased demand for chlorine in the water due to pollution, but on a few occasions this has been the case and the cut-outs have saved the situation.

The normal method of applying the strong chlorine solution at filtration stations is to use the high pressure delivery side of the pumps to supply water to the chlorination instruments, and where two stage pumping is used at well stations the high lift pump pressure injects the chlorine solution against the well pump head. Where single stage pumping is used supplementary solution pumps have to be used to force the strong chlorine solution into the contact tank. These pumps are installed in duplicate to ensure continuity of treatment.

Automatic controls

To a limited extent the MWB treatment is automated, in so far as there are automatic residual controllers to control the level of chlorination, automatic cut-outs, alarm signals, automatic change-over panels, chlorine gas detectors, gas pressure failure alarms and so on. But men must still be available to do tests and attend to the instruments.

At the new Coppermills Works treatment will be operated from a control room with remote slave dials and switchgear. Thus more reliance must fall on the efficiency of residual recorders. These are invaluable for providing permanent records of treatment and for control purposes, but have to be installed with care, regularly serviced and checked, preferably daily, for correctness of reading. Basically a recorder consists of a platinum electrode and a sacrificial copper electrode immersed in a cell through which the water passes. If the electrodes are connected to a sensitive microammeter a current flows momentarily, but ceases due to polarisation. In the presence of depolarising agents, such as chlorine, iodine or bromine, the current is restored and flows in proportion to the concentration and depolarising power of the agent. It is very probable that the effect of a depolarising agent on the electrode system is directly related to the bactericidal efficiency of the agent. Thus chlorine is readily measured and so is iodine but monochloramine has little effect on the cell.

In setting up the instrument it has to be remembered that the rate of flow of test water past the electrodes affects the readings and the time return constant has to be steady otherwise the chlorine residual itself may change.

Finally there is the question of pH adjustment in the cell. It is a matter of experimental fact that this electrode system gives an output at pH 5·5 approximately twice as high as is given at pH 7·5 for any given chlorine strength and it is also a fact that the linearity is not affected by change of pH. The manufacturers advise that a buffer solution, consisting of a mixture of acetic acid and sodium acetate and having a pH of 4·0, should be added to the cell drop wise to produce a pH in the cell of 4·5. The reason given is that this gives *a*) maximum output from the cell and *b*) a consistent output due to the 'blanketing' effect of the buffer. Unfortunately the reaction between chlorine and ammonia to form monochloramine is reversible, and acidification releases chlorine from a preformed monochloramine. Thus the cell content could be quite different from the main stream of water if ammonia were present in the water being treated. The strong buffer solution is unpleasant to handle, the drip feed difficult to control and bacteria thrive in the buffer solution producing white slimes in the cell. However the pH of the Board's many supplies does not vary noticeably from year to year or day to day, even without buffer, so there seems to be no good reason why the buffer should be added at all, provided the cell gives adequate output without its use. No buffer solution has in fact ever been added to recorders in the Kent area wells, and only weak solutions at filtration stations, and the recorders behave excellently without its use.

Relative germicidal efficiencies of chlorine and chlorine compounds

Until recently the expression 'free' chlorine was applied to forms of chlorine, uncombined with ammonia, which exerted a rapid kill on vegetative bacteria. It now seems that if the index is taken as the inactivation of enteroviruses, which appear to be more resistant to chlorine than coliform organisms, then the activity of hypochlorous acid is found to exceed greatly that of hypochlorite ion. Hypochlorous acid is a weak acid and dissociates thus:

$HOCl \rightleftharpoons H^+ + OCl^-$

and unfortunately the dissociation into the relatively inactive OCl^- takes place mainly in the pH range 7 to 8. Thus at pH 7·0 the ratio is 70 per cent HOCl 30 per cent OCl^- and at pH 8·0 it is 20 per cent HOCl and 80 per cent OCl^-. This means that the well water disinfection at pH 7·0 is carried out with 70 per cent of active HOCl present, while at pH 7·7, the pH of river filtered waters, only 35 per cent of HOCl is present in the actual 'free' chlorine dose.

Other uses for chlorine

Mussel and weed control

Prechlorination is still occasionally used, but for specific needs other than disinfection. As stored water leaves the reservoirs freshwater mussels tend to colonise the outlet shafts and the large mains conveying the water to the filtration stations. This is discouraged by intermittent chlorination during their breeding season, *i.e.* from the middle of April till the end of August, a system devised by Greenshields and Ridley.[6] For three periods of 14 days, spread over three months, chlorine is applied to the outlet towers of the largest reservoirs, the dose starting at 0·2 p.p.m. and increasing finally to produce a residual of 0·5 p.p.m. chlorine after contact for 5 min. Before the water enters the filtration stations it is dechlorinated with sulphur dioxide to make sure that no chlorine goes on to the filter beds. The penalty for not applying this treatment is to have the inside of the mains eventually blocked by layer upon layer of mussels.

The New River, which conveys 40mgal./d (184,000 m^3/d) of mixed River Lee and well water to the filtration stations at Stoke Newington and Hornsey, is only 6ft (2m) in depth and slow flowing. The growth of weeds is now checked by chlorination treatment in the summer, instead of by hand cutting which is very expensive. 20mile (32km) out of a total of 24mile (38km) is treated by application of up to 2 p.p.m. at Rye Common and a further addition of up to 1 p.p.m. at Hoe Lane, which is about halfway to Stoke Newington. There is now evidence that the weeds are beginning to give up the struggle. The water flows into two shallow reservoirs at Stoke Newington and here periodic small doses of chlorine are added to prevent the growth of diatoms, particularly *Stephenodiscus hantzschii*.[7] The river weed treatment and reservoir treatment are carefully coordinated.

As might be expected, the mussel and weed treatment chlorination both reduce the numbers of coliforms in the water, but there is no marked change in the dose for terminal disinfection. Much of the prefiltration chlorine is absorbed by organic matter which is removed during filtration and so the water can still absorb more chlorine.

Disinfection of mains

It is the practice of the Board to disinfect all new mains and all repaired mains at a level of 20 p.p.m. of chlorine, to leave water in the main for 2h, empty to waste, refill with district water and to test the water before putting the main back into supply. Mains of 10in (25cm) and over are treated by mobile chlorination units with attendants responsible to supervisors. The units carry gas control instruments and either 4 or 6 × 70 lb (32kg) chlorine cylinders and can inject 500 lb (227kg) of chlorine/d against a head of 500ft (152m). The vehicles are equipped with radio telephone and carry hoses, standpipes and so on, and petrol engine driven pumps, *i.e.* they are mobile chlorine houses requiring only water to operate.

Small mains are treated by 'injectors', small instruments[8] of the Board's patent design which use strong hypochlorite as the source of chlorine and are operated by the distribution engineer's district staff. A total of 76 of these are in use. The injector has to be connected to a live filling hydrant and the 10 per cent hypochlorite is first diluted automatically, by way of an elbow and orifice plate which acts as a venturi, to give a solution of 80 p.p.m. Final dilution is to 20 p.p.m. by way of a venturi constriction in the main water flow. The final strength of solution varies between 20 and 30 p.p.m. depending upon the head of back pressure. All units are equipped with residual chlorine test kits.

Both the gas and the hypochlorite units can be used to dechlorinate the heavily chlorinated water if necessary. The gas units use sulphur dioxide cylinders and dechlorinate the chlorinated water at the point of discharge. With the injector, the hypochlorite is replaced by thiosulphate solution and the chlorinated water recycled through the injector to be dechlorinated. Discharge of heavily chlorinated water to rivers through surface water drains must be avoided in order to prevent fish mortality.

The mobile gas units are also used to disinfect large structures such as contact tanks, service reservoirs, expansion and surge tanks, pump suction chambers, general wells and so on. After work has taken place therein the njector is particularly useful for spraying roofs of enclosed structures, as the spray chlorine dose can be reduced by diluting the hypochlorite and too hazardous an atmosphere is avoided. All persons who deal in any way with chlorine have their own gas masks and resuscitation apparatus is available, all of which is regularly inspected by the first aid officer.

At times the underdrains of filter beds are chlorinated by back filling to the level of the bottom of the fine sand. This is particularly the case at Hornsey filtration station where iron bacteria have taken hold and chlorination must be used to maintain a degree of control.

During the winter months water from some of the Board's works is used to augment underground supplies by means of recharge. The water added is first filtered and chlorinated before being discharged down the well shafts.

Hypochlorite solution is used for swabbing down small jobs on pumps and so on, which cannot conveniently be carried out by other means. Bowls of hypochlorite solution are also used for disinfecting rubber boots before they are worn in the contact tanks and so on, for inspection and other purposes.

Chlorine and enteroviruses

Reference has already been made to enteroviruses. It is now essential that viruses, as well as harmful and undesirable bacteria, should be absent from treated waters for domestic consumption. The Board is fortunate in having a virology unit as part of its water examination department where the staff concerned have been testing the Board's supplies and treated waters for enteroviruses for the past ten years. Viruses can be regularly isolated from the rivers at the Board's points of abstraction, but they have never been found even in very large volumes (of almost 20 litre) of the treated water. This information strongly supports the view that the purification processes of storage, dual filtration and free residual chlorination used by the Board are effective for the elimination of viruses from water before it is pumped into supply. It is believed that each of the three processes, as carried out at the Board's works, contributes to the ultimate complete removal of viruses from the water but the relative importance of each step is a matter for future investigation. The virology unit has recently been expanded to carry out further research on enteroviruses in water supplies, particularly in relation to their removal.

From the point of view of chlorination it is generally accepted that combined chlorine (chloramines) is not effective as a virus destructor but that free chlorine together with an adequate contact period can be effective. Exact figures for the free chlorine concentration that must be present and the time period required cannot yet be laid down because so many factors, such as pH, temperature and local circumstances, are involved. But a fair estimate in general terms would be to say a free chlorine residue of 0·5mg/l after a contact period of 30min to 1h would be satisfactory.[9,10]

References

1 Griffin, A. E., *Jour. A.W.W.A.*, 1940, **32**, 1187
2 Fair, G. M., *ibid.*, 1948, **40**, 1051
3 Tarras, M. J., *ibid.*, 1953, **45**, 47
4 Palin, A. T., *Water & Water Eng.*, 1950, **54**, 151, 189, 249
5 Byles, D. B., *Proc. Soc. W.T. & E.*, 1955, **4**, 69
6 Greenshields, F. & Ridley, J. E., *J. Inst. W.E.*, 1957, **11**, 300
7 Windle Taylor, E., *Proc. Int. Ass. Limnology.*, 1955, **12**, 671
8 Bell, R. J., *Trans. I.W.E.*, 1945, **50**, 284
9 Weidenkopf, S. J., *Virology*, 1958, **5**, 50
10 Poynter, S. F. B., *Proc. Soc. W.T. & E.*, 1968, **17**, 187

TREATMENT OF WASTES FROM A SOLE LEATHER TANNERY

J. David Eye and Lawrence Liu

The tanning industry long has been recognized as a major contributor to water pollution because of the highly pollutional nature of the waterborne components of untreated tannery effluents. The overall volume of tannery wastes in the U. S., however, amounts to only about 16 bil gal/yr (61 mil cu m/yr), with the sole leather tanneries contributing approximately 10 percent of this volume. On a national basis, therefore, the wastes from sole leather tanneries are relatively insignificant, whereas on a local or regional basis they often are of major concern.

While studies on the treatment of tannery wastes were started in the U. S. about 1850, few full-scale treatment plants have been built for handling tannery wastes. A detailed investigation of the tanning industry in the U. S. in 1965–66 revealed that, while a number of tanneries were served by various treatment procedures, no tannery had acquired a treatment system that was completely satisfactory. Operational data gathered during the survey indicated that most of the systems had been designed improperly from the standpoint of the effects of specific constituents of tannery wastes on conventional waste treatment processes.

However, the tanning industry recognized the need for finding technically feasible and economical means of waste treatment that might be employed throughout the industry. During 1966–67, several laboratory/pilot plant studies on the treatment of tannery wastes from a sole leather tannery were carried out at the International Shoe Company Tannery, located at Marlinton, W. Va. These studies were sponsored jointly by the Tanners' Council of America, the Water Resources Commission of West Virginia, and the University of Cincinnati. The data derived from the pilot-plant studies formed the basis for a demonstration project supported by the Federal Water Quality Administration (FWQA) (now Environmental Protection Agency–Water Quality Office) for the design, construction, and operation of a full-scale waste treatment system at the Marlinton tannery.

Tanning Processes and Wastes

About 800 heavy steer hides are processed into sole leather at the Marlinton plant on each of the five working days per week. While many individual steps are required for converting the hides into leather, they can be grouped under two major operations, beamhouse and tan yard. In the beamhouse operations the hides are prepared for tanning, and in the tan yard the skins are converted into sole leather. Salt-cured hides are used.

The beamhouse operations include an initial wash and soak to remove curing salt, extraneous dirt, blood, and manure and to soften the hides. After the hides are washed and soaked, they are immersed in a lime-sulfide solution in still vats or pits. The lime and sulfides dissolve the unwanted hide substance and loosen the hair. After the hides are removed from the lime vats, they are rinsed to remove excess chemicals, unhaired, fleshed, and sent on to the bating process. Bating consists of washing the hides in a solution containing wetting agents, enzymes, and ammonium salts to remove excess lime and to prepare further the hides for tanning.

In the tanning operation the hides are gently rocked for a period of several weeks in a solution made from vegetable extracts. The vegetable extracts react with the collagen fibers to produce leather. After the tanning step, the leather is run through a bleaching process to remove excess tannin and to give the desired color control. Final finishing operations for sole leather are mainly mechanical in nature and are designed to impart specific characteristics to the leather.

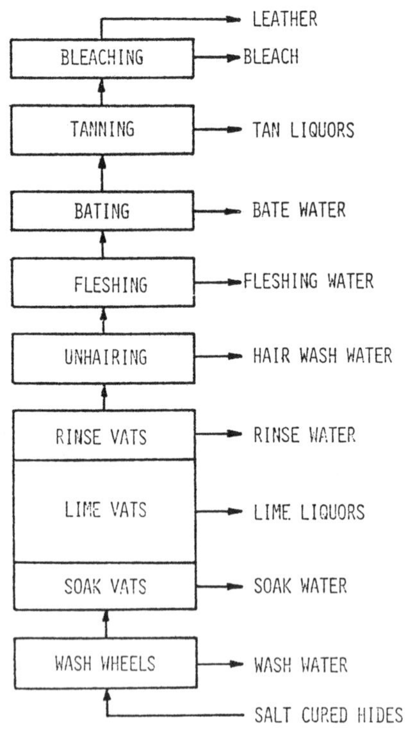

FIGURE 1.—Sources of major tannery wastes.

The beamhouse wastes constitute about 70 percent of the total waste volume and contain a large fraction of the pollutional load. Because the beamhouse operations are carried out on a batch basis, most of the pollutants initially are contained in relatively small volumes of waste. The pH of the total beamhouse wastes ranges from 11.5 to 12.5.

The spent tan liquors are extremely high in color and chemical oxygen demand (COD) and moderately high in biochemical oxygen demand (BOD). The pH averages about 4.5, and the acidity is sufficient to reduce the pH of the total tannery wastes to about 9.5 when all waste streams are mixed.

The major waste fractions stemming from the tanning operations at the Marlinton plant are illustrated in Figure 1. Some of the more important characteristics of the individual waste streams are listed in Table I.

As shown in Table I, the lime vat, rinse vat, and hair washer waters contain moderate to high concentrations of COD and suspended solids (SS) mostly $Ca(OH)_2$ and have a high pH, yet they make up only 32 percent of the beamhouse wastewater volume. The wash, soak, and bating waters represent 64 percent of the waste volume, but

TABLE I.—Characteristics of Tannery Waste Fractions

Waste Fraction	Flow (gpd)	COD (mg/l)	SS (mg/l)	pH
Wash water	25,000	2,100	1,300	6.8
Soak water	10,000	2,200	1,000	7.8
Lime water	10,000	11,900	30,300	12.3
Rinse water	20,000	2,500	4,900	12.3
Hair water	15,000	2,500	3,100	12.3
Fleshing water	5,000	3,600	4,900	12.3
Bate water	55,000	1,700	1,000	9.0
Spent tan liquors	60,000	10,000	500	4.5

Note: Gpd × 3.785 = l/day.

are moderate to low in COD and SS and near neutral in pH.

When the concentrated waste fractions are mixed with the large volumes of wash waters and other only slightly contaminated wastes, the resulting or final beamhouse waste stream is large in volume and still grossly polluted. For example, lime, while soluble to a rather limited extent in water, will continue to dissolve as the liquors containing high concentrations of suspended lime are mixed with non-lime-bearing wastes, thereby increasing the hardness, alkalinity, and pH of the combined waste streams. The high pH resulting from the lime also precludes any form of biological treatment for reduction of the BOD of the wastes unless the waste is partially neutralized.

Neutralization of the excess lime with acid is costly and leaves the waste with a high calcium content. Also, neutralization with acid must be controlled carefully because of the danger of liberating hydrogen sulfide from the sulfides contained in the waste.

The spent tan liquors, when mixed with the beamhouse waste streams containing lime, yield a voluminous precipitate which is difficult to separate from the liquid phase. In addition, the colored compounds present in the spent tan liquors are sufficiently concentrated to impart an extremely intense color to the total tannery wastes.

Research Plan

In general, small volumes of concentrated wastes are easier and more economical to treat than large volumes of a more dilute waste. Also, in many cases it is easier to remove the pollutants from the individual waste streams than from the combined wastes.

Ceamis,[1] Jansky,[2] Rosenthal,[3] Guerree,[4] and Eye and Graef [5] have shown that the combined tannery wastes are amenable to biological treatment if the suspended lime is removed as a pretreatment measure.

It was determined, therefore, that the basic approach to be used for the Marlinton project would be that of separating the waste streams, removing the excess suspended lime from the beamhouse wastes, and blending all waste streams for final treatment by biological means. Pilot-plant studies were used to provide design data for the full-scale system. In general, each unit or treatment process was constructed and evaluated before the next downstream unit was constructed. This "step-by-step" procedure provided a desired degree of flexibility to the design of the total system and allowed for easy modification when changes had to be made in the basic plan.

Removal of Suspended Lime

Ceamis,[1] Jansky,[2] and Domanski [6] investigated the use of iron salts as coagulants for the lime-bearing waste fractions. Sproul,[7] Scholz,[8] and Eye and Liu [9] have reported on the effectiveness of polyelectrolytes in tannery waste treatment.

Laboratory and pilot-plant studies indicated that an anionic polyelectrolyte was highly effective in clarifying the lime-bearing beamhouse effluents. Polyelectrolyte * dosages from 0 to 50 mg/l were investigated. Little improvement in the rate or degree of clarification was noted at dosages above 10 mg/l. It also was found that 5 mg/l gave approximately the same removal of SS as 10 mg/l. The rate of settling at the 10-mg/l dosage, however, was twice as great as for 5 mg/l, and a dosage of 10 mg/l was considered optimum for design purposes.

Typical results obtained from the pilot-plant studies at a polyelectrolyte dosage of 10 mg/l are presented in Figure 2. The data obtained from the pilot clarifier indicated that SS removals on the order of 90 percent and 70 percent could be achieved at overflow rates of 2,000 gpd/sq ft and 3,000

* A-10 Polyelectrolyte, manufactured by Rohm & Haas, Philadelphia, Pa.

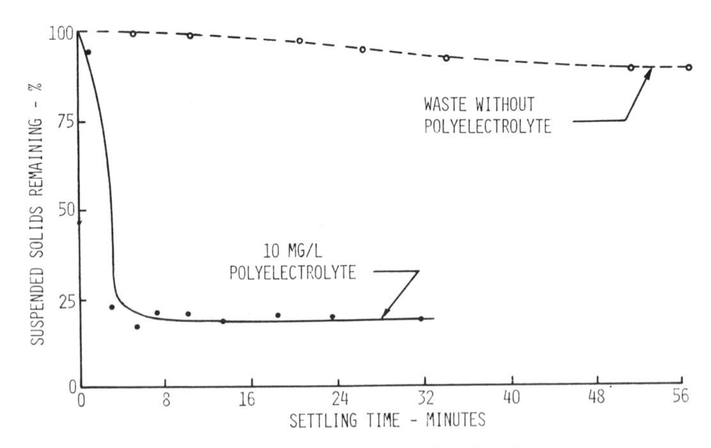

FIGURE 2.—Settling curves for lime-bearing wastes.

gpd/sq ft (81.6 and 122.4 cu m/day/sq m), respectively.

The success achieved in the pilot-plant studies prompted a decision to design and construct a full-scale system to clarify the total lime-bearing wastes discharged from the beamhouse. A preliminary plan for separating the waste fractions was developed, and the clarification unit complete with polyelectrolyte-feeding equipment was designed.

Full-Scale Clarification System

Design

The layout of the process units and the sewer system at the start of the project is illustrated in Figure 3. The wastes discharged from the 10 initial soak vats and the 30 lime-sulfide and rinse vats were carried in a common sewer located beneath the battery of vats. Construction of an auxiliary sewer underground to serve the soak vats independently of the lime vats

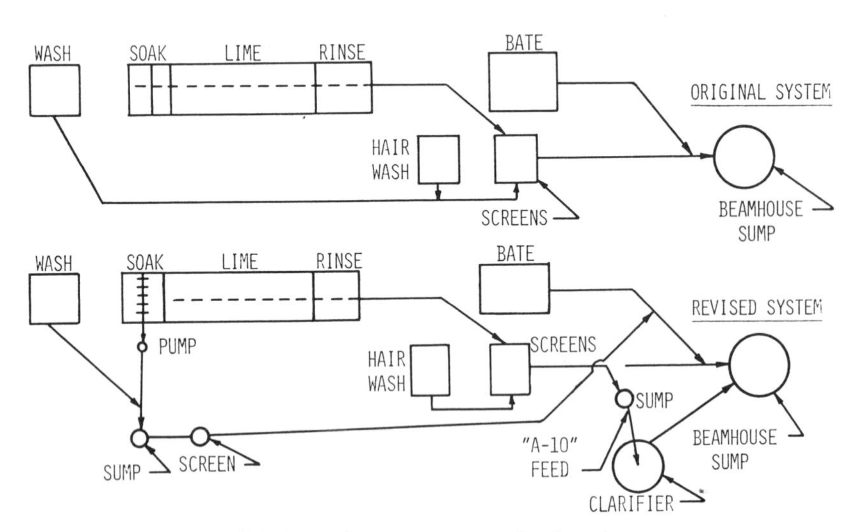

FIGURE 3.—Sewer system serving beamhouse.

would have been extremely difficult and expensive. It was decided, therefore, to empty the soak vats by use of a pump and an overhead piping system.

The revised flow diagram for the beamhouse sewerage system also is illustrated in Figure 3. Because the lime-bearing wastes are discharged intermittently over an 8-hr period on each working day, it was determined that a holding sump would be advantageous from the standpoint of clarifier operation. A sump with a capacity of about 2,000 gal (7.57 cu m) was constructed near the end of the lime liquor discharge channel and the lime liquors diverted to the sump. A float-actuated pump was installed in the sump to pump the lime-bearing wastes to the clarifier. Provision was made to inject the polyelectrolyte solution into the discharge line from the sump pump by means of a small gear pump that operates only when the main pump is running. The discharges of the sump pump and the chemical feed pump can be adjusted to accommodate flows up to more than 100,000 gal (378.5 cu m) in an 8-hr period.

The clarifier was designed to provide a detention time of 30 min and an overflow rate of 2,000 gpd/sq ft (81.6 cu m/day/sq m) at a feed rate of 150 gpm (9.47 l/sec). A cylindrical steel tank 12 ft (3.66 m) in diam and 11 ft (3.35 m) deep was selected. These dimensions met the design requirements and, more important, permitted the tank to be fabricated at the factory and transported to the site by truck. The cost of factory fabrication was approximately 50 percent less than that for field construction of a similar unit. The details of the clarification system are shown in Figure 4.

Clarifier Performance

The data shown in Figure 5 illustrate the effectiveness of removal of SS at two overflow rates for a polyelectrolyte dosage of about 10 mg/l. In general, the removal of SS exceeded 80 percent even at the higher overflow rates. Because the objective of clarification was to remove suspended lime, the data for removal of fixed SS give a more realistic measure of the efficiency of clarification. During one period of 2 months duration, the daily removal of fixed SS averaged 93 percent. The overflow rate and polyelectrolyte dosage during the period were 1,600 to 2,000 gpd/sq ft (65.3 to 81.6 cu m/day/sq m) and 8 to 12 mg/l, respectively.

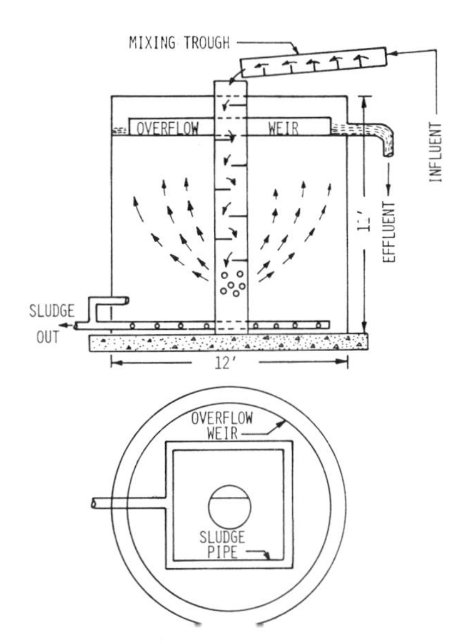

FIGURE 4.—Upflow clarifier details.

The effect of polyelectrolyte dosage on fixed SS removal is illustrated in Figure 6. The average removal was about 8 percent greater at polyelectrolyte dosages of 7 to 11 mg/l than at 4 to 7 mg/l for the same range of overflow rates.

The dosage of polyelectrolyte used in this system was higher than normally considered economical in water and waste treatment, but only the lime-bearing waste fractions, which represented about 30 percent of the beamhouse flow, required treatment, and the

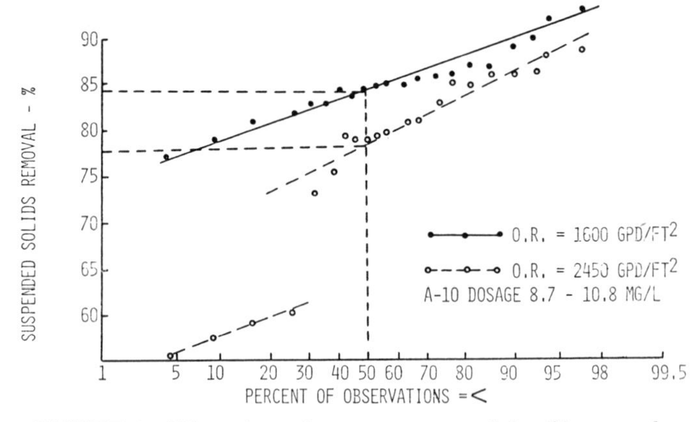

FIGURE 5.—Effect of overflow rate on suspended solids removal. (Gpd/sq ft × 0.0408 = cu m/day/sq m.)

actual weight of polyelectrolyte used each day was relatively small.

Close observation of the operating characteristics of the clarifier revealed that the removal of SS was correlated closely with the depth of sludge maintained in the unit. When the accumulated sludge depth became too great, solids were washed from the unit, thereby producing decreased removal efficiencies. Optimum removal usually was achieved when the sludge layer was maintained at 4 to 6 ft (1.22 to 1.83 m) below the water surface. The depth of sludge was maintained in the desired range by frequent pumping of the accumulated sludge from the bottom of the clarifier. When the depth of accumulated sludge was less than about 2 ft (0.61 m), a considerable volume of water was withdrawn during sludge pumping with a resulting decrease in the solids content of the sludge.

The sludge as withdrawn from the clarifier had a solids content ranging from 8 to 30 percent with an average of 14 percent. The volume of sludge

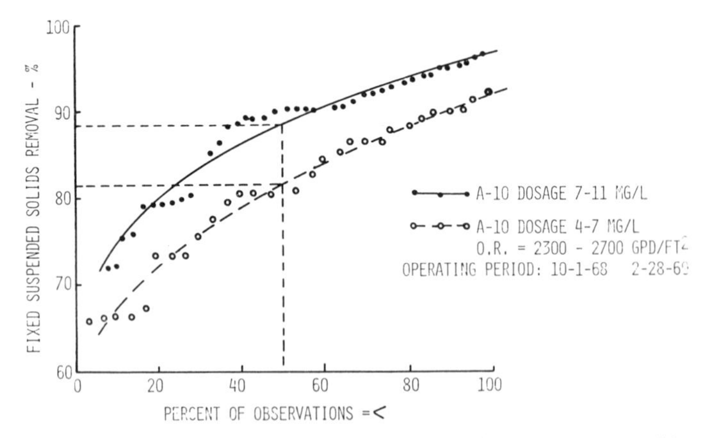

FIGURE 6.—Effect of polyelectrolyte dose on fixed suspended solids removal. (Gpd/sq ft × 0.0408 = cu m/day/sq m.)

produced amounted to about 3 percent of the volume of lime-bearing wastes treated. The sludge dried readily when placed on drying beds or spread on the surface of the ground. The results obtained from operating the clarifier for more than 1 yr demonstrated the feasibility of utilizing polyelectrolytes and settling for removing the excess lime as a pretreatment procedure.

Biological Treatment

Separation and pretreatment of the various waste fractions, while effective in removing the inert SS, effected only a limited reduction in the total organics contained in the wastes. The BOD (5-day, 20°C) of the pretreated and blended waste streams from the beamhouse ranged from 1,000 to 1,500 mg/l, and the COD from 2,000 to 3,000 mg/l. The total tannery wastes after pretreatment and blending had a BOD of 1,500 to 3,000 mg/l and a COD of 4,000 to 8,000 mg/l. The total Kjeldahl nitrogen concentrations in the beamhouse wastes and the total tannery waste averaged about 200 and 150 mg/l, respectively. The total sulfide concentration ranged from 20 to 50 mg/l.

It was determined that the organic content (5-day, 20°C BOD) of the total wastes would be reduced by 80 to 90 percent before the wastes would be discharged to the receiving stream. The economic position of the sole leather industry dictated that the treatment system selected for reducing the organics have a low capital and maintenance cost and be relatively easy to operate and maintain. Another important consideration in the selection of a system for removing the organics was that the production of sludge be minimal so that extensive sludge-handling facilities would not be required.

A combination of anaerobic and aerobic biological units seemed to meet the basic requirements established for the system, particularly if they could be combined in a lagoon or series of lagoons. Ivanof[10] and Toyoda[11] reported on the successful treatment of sole leather tannery wastes by anaerobic means. Gates and Lin[12] conducted laboratory and pilot-plant studies on a stratified anaerobic-aerobic lagoon process and found it applicable to treating tannery wastes.

A decision was made to explore the feasibility of combining an anaerobic and an aerobic biological process in a deep lagoon to achieve the desired removal of organics. A deep lagoon equipped with floating aerators arranged to aerate only the upper zone of the wastes being treated offered the potential advantages of: (*a*) low construction cost where soil conditions were favorable, (*b*) small land area requirements, (*c*) low volume of sludge accumulated, (*d*) reduced air requirements for the aerobic system because some organics would be eliminated in the anaerobic zone; and (*e*) heat conservation during winter operation. The large volume of wastes undergoing biological breakdown also would tend to protect the biological system against shock loads, which are always possible from batch operations in a sole leather tannery.

This concept was evaluated in pilot-plant studies, and the data derived from the pilot studies provided a basis for the design of a full-scale anaerobic-aerobic lagoon system.

Anaerobic-Aerobic Lagoon System

After removal of the suspended lime, the clarified lime waste was mixed with the other beamhouse waste fractions for treatment in an anaerobic-aerobic lagoon system that provided a detention time of about 4 days. The wastes were pumped to a lagoon that was approximately 12 ft (3.66 m) deep and equipped with three floating aerators having a combined power rating of 30 hp (22.4 kw).

The pH of the lagoon was reduced to approximately 9.0 by adding sulfuric acid. The aerators were started, and within a few days biological activity became apparent. The BOD and COD of the treated wastes decreased continuously, and pronounced odors emanated from the lagoon. Neutralization of the influent wastes to the lagoon was discontinued as the pH stabilized at 7.8 to 8.0. Odors continued to be a problem and continued odor control was necessary because of the close proximity of the lagoons to residences in the town. At no time, however, was there any evidence of hydrogen sulfide being released from the operating lagoon.

The effluent from the biological system was discharged into a second lagoon before being discharged to the receiving stream. Within about 1 wk after start-up of the anaerobic-aerobic lagoon, a dense growth of algae appeared in the second lagoon. Over a period of several weeks the algae became so dense that the dissolved oxygen (DO) was completely depleted during night time, and hydrogen sulfide was released from the bottom deposits. Thus, while hydrogen sulfide was no problem in the operating lagoon, it became a serious problem in the lagoon that received the treated effluent. Some 10 houses adjacent to the second lagoon showed severe darkening of the paint, and reimbursement of the owners by the insurance company was necessary.

The BOD reduction ranged from 80 to 85 percent while the biological system was operating. The total Kjeldahl nitrogen content was reduced by about 50 percent, but no reduction in sulfides was observed. No measureable settleable solids ever appeared in the effluent although the SS averaged about 200 mg/l. It was apparent, therefore, that the beamhouse wastes were amenable to biological degradation.

The odors were eliminated by decreasing the flow and concomitantly the organic load to the anaerobic-aerobic lagoon. The reduction in flow was achieved by diverting a portion of the wastes through an existing lagoon before discharge to the receiving stream. With the onset of cooler weather in the autumn of 1968, foaming became a major problem. On many nights, a stable foam layer 5 to 6 ft (1.52 to 1.83 m) thick would completely cover the lagoon. All efforts to control the foam failed, and a decision was made to construct new lagoons at a site about one mile downstream. The new lagoons (Figure 7) provided a surface area of about 60,000 sq ft (5,574 sq m) and a volumetric capacity of about 2.3 mil gal (8,706 cu m). The new lagoons had a depth of only 6 ft (1.83 m) and provided a detention time of about 16 days for the total flow.

In May 1969, the lagoons were filled with clarified beamhouse wastes (the lagoons used in 1968 served to clarify the lime-bearing wastes), and the aerators were started. No effort was made to reduce the pH of the wastes before start-up of the biological system, and the wastes were not "seeded" with domestic wastewater. Within 2 days the pH had fallen to about 9.5, and it was apparent that biological activity was under way.

The units were operated for approximately 2 months on beamhouse wastes. Odors were apparent in the vicinity of the lagoons, and severe foaming occurred intermittently. On July 12 and 13 about 500,000 gal (4,353 cu m) of spent tan liquor were added to the system. Tan liquors then were added intermittently from July 13 through August 20. The BOD of the effluent increased for several days after the tan liquor was first added, then decreased to about the same level as had been observed when only beamhouse wastes were being treated. The odors disappeared completely, and foaming was not nearly so severe after tan liquor was added to the lagoons.

Starting August 22, all of the tan liquors were added to the system, thereby increasing both the influent and

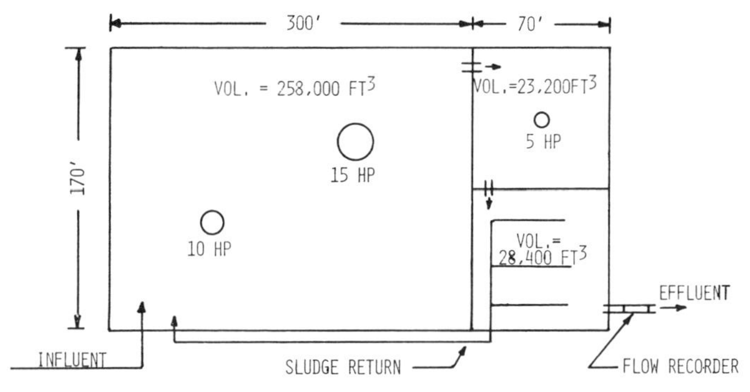

FIGURE 7.—Layout and approximate dimensions of lagoon system. (Cu ft × 0.028 = cu m; hp × 0.75 = kw; ft × 0.305 = m.)

effluent BOD values. The spent tan liquors, however, were not mixed with the beamhouse wastes before introduction to the biological system. Each waste fraction was pumped separately to the treatment site through the same pipeline. The BOD and COD loading in the biological system has averaged about 3 to 5 lb/day/1,000 cu ft (48 to 80 g/day/cu m) (5-day, 20°C BOD) and 6 to 10 lb/day/1,000 cu ft (96 to 160 g/day/cu m), respectively.

A detailed survey of the lagoons after 4 months of operation showed that the upper two-thirds of depth of the aerated lagoons were aerobic and the bottom third anaerobic. The accumulated sludge depth averaged less than 2 in. (5.1 cm) and there was evidence of considerable anaerobic decomposition of the deposited material. No settleable solids were present in the effluent although the SS concentration showed a slow but steady increase from 50 mg/l at the start of operation up to about 200 mg/l.

The data presented in Figure 8 illustrate the operating conditions and the results obtained for the period May 19–October 20, 1969. While the 5-day, 20°C BOD is widely used to characterize waste loads, long-term BOD's were run periodically (Figure 9). The data show that for the effluent, the 5-day values represent a major part of the ultimate BOD. Some of the long-term values for the influent were considerably higher than the 5-day values. In the design of aeration equipment for a lagoon with a detention time greatly in excess of 5 days, the long-term BOD values must be given serious consideration. Another factor to consider in the design of the aeration equipment is the solubility of oxygen in the waste to be treated. Figure 10 illustrates the fact that tannery wastes do not have as high a saturation value as pure water, and, furthermore, they exert a rapid uptake of oxygen in the absence of active aeration. It seems that much of the oxygen uptake is chemical rather than biological in nature because of the fact that sterilized samples gave a similar pattern of DO versus time.

The pH of the operating lagoons has remained stable at 7.8 to 8.1 throughout the operating period. It is apparent, therefore, that neutralization of the waste fractions is not necessary from the standpoint of biological degradation of the wastes. No reduction in total sulfides has been noted, and there has been no evidence of hydrogen sulfide generation even though the pH normally is slightly less than 8.0.

Effect of Effluent on the Receiving Stream

The effluent became highly colored soon after spent tan liquors were added

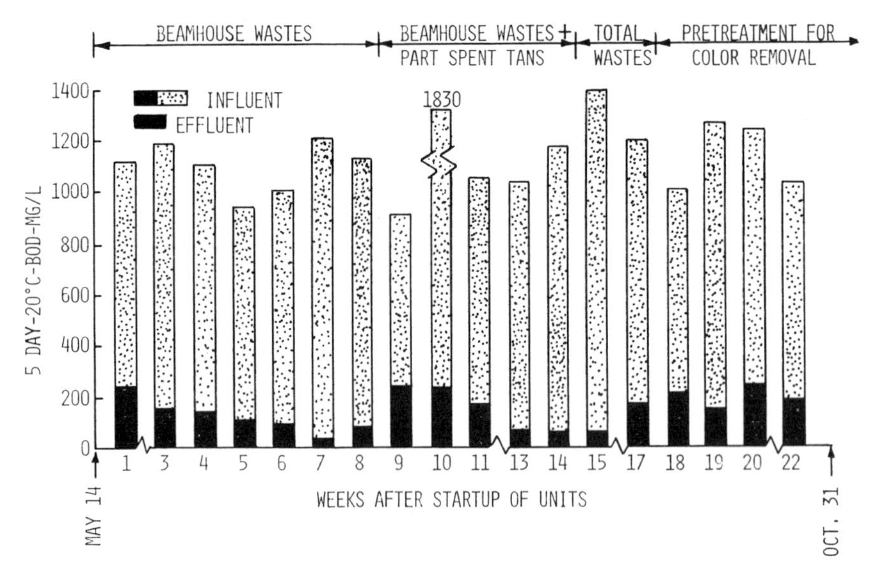

FIGURE 8.—Reduction in biochemical oxygen demand in biological treatment system.

to the biological units. A survey of the receiving stream revealed that the effluent reduced the DO in a narrow segment of the stream immediately below the point of discharge. This segment of the stream also was highly colored, but aquatic life was abundant. The discoloration persisted for at least 1 mile (1.6 km) downstream. It is believed that dispersal of the wastes across the entire width of the stream would reduce the color to a more acceptable range, although the entire stream would be slightly discolored during periods of low stream flow.

Removal of Color

After having determined that the total tannery wastes could be treated effectively in an anaerobic-aerobic biological system, it was decided that efforts should be made to remove the residual color from the lagoon effluent. Detailed laboratory and pilot-plant studies demonstrated that the residual color in the lagoon effluent could be

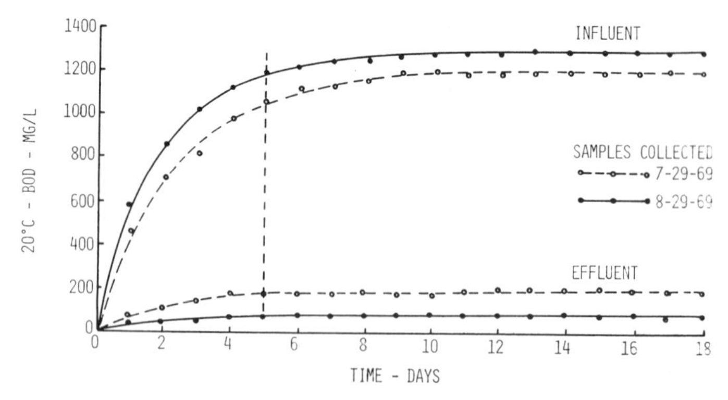

FIGURE 9.—Long-term biochemical oxygen demand values.

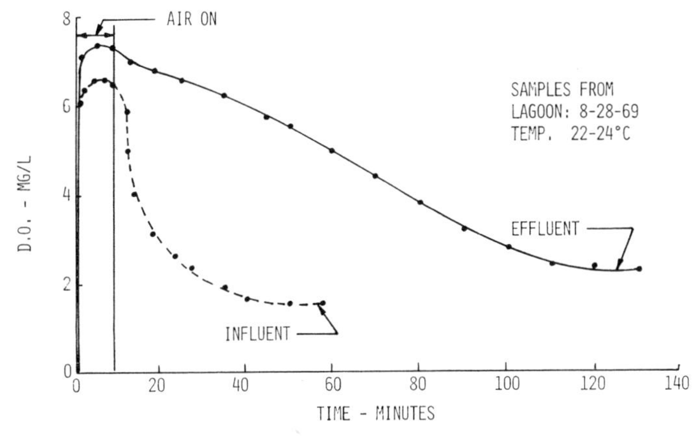

FIGURE 10.—Oxygen buildup and uptake.

removed effectively by adding lime to bring the pH to about 12.0. The addition of an anionic polyelectrolyte † at a dosage of 2 to 5 mg/l produced rapid settling of the precipitated color compounds leaving the effluent with only a pale yellow tinge. A reduction in color of at least 90 percent was achieved (estimated by dilution with river water) and the volume of sludge produced was small. The dosage of lime required to increase the pH of the effluent to 12.0, however, was in excess of 2,000 mg/l. This fact, coupled with the necessity of reducing the pH to 10.0 or less before final discharge, rendered the process uneconomical.

The studies then were directed toward precipitating the color-forming compounds before biological treatment. It was found that by mixing the spent tan liquors with the highly alkaline beamhouse waste fractions, the colored materials were precipitated when the pH was maintained above 11.5. In the laboratory and pilot-plant studies, the mixture of the two waste fractions produced a large volume of sludge that settled poorly. Efforts to overcome the sludge problem by use of polyelectrolytes (in a reasonable dosage range) were unsuccessful.

† NALCO-675 manufactured by Nalco Chemical Company, Chicago, Ill.

It was decided, however, to conduct a full-scale experiment in mixing the two wastes before discharging them to the biological treatment unit. The two waste fractions were mixed in a small lagoon and allowed to pass through several larger lagoons before reaching the biological unit. The reduction in color was dramatic and the wastes, if discharged directly to the receiving stream, would present no color problem. The resulting precipitates settled rapidly and appeared to compact readily. This finding was quite surprising in light of the laboratory and pilot-plant experience.

Operation of the color removal system over a period of months has shown that the color will not be removed effectively unless the pH on initial mixing of the two waste fractions is above 11.5. It also has been noted that the precipitated materials will redissolve if the pH of the wastes in contact with the precipitates drops below about 10.5.

A mixing unit in which the pH is automatically controlled by the addition of an excess of highly alkaline wastewater coupled with a clarification basin from which the sludge can be removed continually or frequently probably will be required. No data are available at present for formulating criteria for the design of a clarification and sludge-handling system for removal of color from spent tan liquors.

Disinfection of Effluent

The tannery wastes as discharged from the lagoon system had a high bacterial count. Disinfection of the effluent with chlorine proved difficult because of the high concentrations of ammonia and sulfides. Chlorine dosages of 30 to 50 mg/l with a contact time of 15 to 25 min were required to give satisfactory reduction in total bacteria and coliform counts.

Conclusions

The data derived from the pilot-plant and full-scale treatment studies, which extended over a period of approximately 3 yr, lead to the following specific conclusions:

1. A detailed study of the total tanning operations is a required first step in formulating a feasible waste treatment procedure. Specifically, the sources of all wastes must be identified, and each waste stream must be completely characterized. The volume, discharge pattern, and constituents of each waste fraction must be determined accurately and related to specific tanning operations.

2. A waste reduction program through conservation, reuse, and process changes is feasible for a sole leather tannery. Such a program can be effective only if the plant operating personnel are fully informed of the objectives to be achieved and the role that they play in the total plan.

3. About 70 percent of the total pollutional load discharged from a sole leather tannery initially is contained in three or four waste streams that comprise only about 30 percent of the total volume of wastes discharged. Segregation and pretreatment of the individual waste fractions, therefore, is necessary if an economical waste treatment procedure is to be achieved.

4. Excess hair, fleshings, and grease should be removed from the waste streams at an early point in the waste management procedure, as these materials clog pumps and generally interfere with any mechanical handling of the wastes.

5. In the full-scale system, an anionic polyelectrolyte at a dosage of 10 mg/l provided optimum removal of the suspended lime particles from the lime waters. Removal efficiencies in excess of 90 percent were achieved routinely at clarifier overflow rates of 1,600 gpd/sq ft (65.3 cu m/day/sq m) of clarifier surface area. Even at overflow rates of 2,000 to 2,500 gpd/sq ft (81.6 to 102 cu m/day/sq m), removal efficiencies of 80 to 90 percent were quite common.

6. The reduction in BOD through both the pilot and full-scale units ranged from 70 to 90 percent. The addition of spent tan liquors to the biological system adversely affected the overall performance as measured by the effluent BOD and COD values.

7. Foaming of the aerated lagoons occurs periodically and is severe enough to prohibit the location of such a system near residential or commercial areas. High-pressure water jets were effective in controlling the foam when air temperatures were above freezing but could not be used during the winter months.

8. Severe odor problems were encountered when the anaerobic-aerobic lagoons were operated on beamhouse wastes only. The addition of the spent vegetable tan liquors eliminated the odors completely.

9. Large numbers of bacteria are present in the final effluent. Adequate disinfection can be achieved with chlorine at a dosage of about 30 mg/l with a 15-min contact time. The treated waste exerts an extremely high chlorine demand, but the reaction seems to be sufficiently slow to permit high bacterial kills before the chlorine disappears.

10. The installed cost of the Marlinton system was approximately $40,000. The operating costs are estimated at about $15,000/yr or $0.07/hide for a production level of 800 hides/day.

Acknowledgments

Many individuals and organizations were involved in the total project. Assistance rendered by various individuals from the following organizations is gratefully acknowledged: The Tanners' Council of America; The University of Cincinnati; West Virginia University; The West Virginia Water Resources Commission; The International Shoe Company; and FWQA. This project was supported in part by Grant No. WPD-185 from the FWQA.

References

1. Ceamis, M., "Noxiousness and Purification of Tannery Waste Waters." *Ind. Usoara* (Rom.), **2**, 208 (1955).
2. Jansky, K., "Tannery Waste Water Disposal." *Kozarstvi*, **11**, 327 and 355 (1961).
3. Rosenthal, B. L., "Treatment of Tannery Waste Sewage Mixure on Trickling Filters." *Leather Mfg.*, **74**, 12, 20 (1957).
4. Guerree, H., "Purification of Tannery Waste Water." *Bull. Assn. Franc. Ingr. Chim. Tech. Ind. Cuir Doc. Inferm Centre Tech. Cuir*, **26**, 95 (1964).
5. Eye, J. D., and Graef, S. P., "Pilot Plant Studies on the Treatment of Beamhouse Wastes from a Sole Leather Tannery." *Jour. Amer. Leather Chem. Assn.*, **63**, No. 6 (1968).
6. Domanski, J., "Sedimentation of Suspension in Coagulation of Sewage from Tanning Industry." *Gaz. Wodai Tech. San.*, (Pol.), **38**, 279 (1964).
7. Sproul, O. J., *et al.*, "Extreme Removals of Suspended Solids and BOD in Tannery Wastes by Coagulation with Chrome Dump Liquors." *Proc.* 21st *Ind. Waste Conf.*, Purdue Univ. Ext. Ser. **121**, Vol. L, No. 2 (1966).
8. Scholz, H. G., "Modern Effluent Water Disposal in the Leather Industry—Effects and Cost." Lectures during the 8th Congress of the International Union of Leather Chemists Societies, (1963).
9. Eye, J. D., and Liu, L., "Clarification of Lime Bearing Wastes from a Sole Leather Tannery." *Proc. 24th Ind. Waste Conf.*, Purdue Univ., Ext. Ser. **135** May, 1969.
10. Ivanof, G. I., "Anaerobic Purification of Tannery Waste." *Kozh. Obuvn. Prom.*, **4**, 7, 30 (1962).
11. Toyoda, H., *et al.*, "Studies on the Treatment of Tannery Wastes." *Nihon Hikaku Gijutsu Kyokai-Shi*, **8**, 79 (1963).
12. Gates, W. E., and Lin, S., "Pilot Plant Studies on the Anaerobic Treatment of Tannery Effluents." *Jour. Amer. Leather Chem. Assn.*, **61**, 10 (1966).

THE ECONOMIC FEASIBILITY OF TREATING TEXTILE WASTES IN MUNICIPAL SYSTEMS

Kimrey D. Newlin

With the vigorous pollution abatement activities that have been undertaken by state and federal governments, industries and communities are becoming increasingly concerned with finding an economical means of disposing of their wastes. Many communities have built waste treatment plants and have accepted industrial wastes into these treatment plants. There have been many cases where industrial wastes, including those of textile plants, have reduced the efficiency of the treatment plants and have at times rendered them useless.[1] Therefore there has been increasing concern about the effects of industrial wastes on municipal wastewater treatment plants.

From the review of literature[2] one would assume that if a municipal disposal system were properly designed, its managers should be able to make satisfactory arrangements with textile plants to treat their wastes with some "specified compensation by the industry for doing so. Herein lies the value of any study which provides some indication of the costs with which industry may be expected to be encumbered, and therefore permit comparisons and planning to seek the better alternative, of whether to use public or private means."[3]

In a 1962 report, "Intergovernmental Responsibilities for Water Supply and Sewage Disposal in Metropolitan Areas," the Advisory Commission on Intergovernmental Relations said: "Because of the economies of scale involved, municipal and metropolitan sewer and sewage treatment facilities should be designed to accommodate industrial waste which can be handled without damage to the system."[4]

In another report, "Industrial Incentives for Water Pollution Abatement," the editor supported the view that economies of scale may exist by using one waste treatment facility to treat both domestic and industrial wastes.[5] This report also indicated that the textile firms, by the use of a public treatment facility system, could pass on their responsibility for meeting state pollution control standards to the municipality, thereby placing a burden on the municipality.

As mentioned earlier, limitations exist on the kinds and amounts of industrial pollutants that may be accepted in municipal waste treatment facilities. Present facilities may lack the size and ability to handle large volumes of complex industrial wastes. Some cases have been cited where it was more economical for large industrial waste treatment plants to accept and treat domestic wastes.[5]

For the successful treatment of textile wastes with domestic wastewater, other important factors have to be considered such as types of chemicals found in textile wastes, temperature of textile wastes, and relative volume of textile wastes in proportion to the volume of domestic waste.

Imhoff and Fair [6] stated that most industrial wastes can be treated in municipal treatment plants along with domestic wastes. Several exceptions were made. "Grease and oil, hot liquids (above 95°F (35°C)), gasoline and flammable solvents, concentrated acids and poisonous substances, if present in sufficient amounts, are destructive or dangerous to sewage conduits (corrosion, clogging, explosion, or other damage) and to treatment works and processes. Therefore, municipal authorities should be empowered: (1) to exclude objectionable wastes from the sewage systems; (2) to regulate the rates at which potentially dangerous wastes may be admitted to sewers; (3) to prescribe the degree of pretreatment to which wastes should be submitted before discharge; and (4) to determine how much of the cost of treatment of the mixed domestic and industrial waste waters should properly be paid for by industry." [6]

Masselli *et al.*[1] pointed out that many conditions may cause trouble and that because there is no simple examination or remedial procedure for detecting or preventing early damage, the waste treatment plant operator must make use of all the analytical procedures available to him in order to examine the quantity and the quality of the wastewater entering and leaving his plant daily—or even hourly—to discover changes that might mean potential trouble. He must realize the design capacities of his waste treatment plant and the characteristics of the wastes and have a method of timing discharges from his contributing industries. Thus, when these discharges increase or are likely to cause trouble, the operator can develop effective procedures for handling the situation. Sometimes he may be able to develop additional analytical procedures to serve as specific indicators for a certain type of waste. Masselli *et al.*[1] also noted the effects of textile and industrial wastes on wastewater treatment plants and described methods that might alleviate their effects. They noted that this is very important, for in the future the discharge of industrial wastes into wastewater treatment systems will increase and present a challenge to the plant designer or operator who must handle them wisely. Further research is needed to denote the types and amounts of wastes that may be permitted in wastewater systems and to determine how to reduce their effects. Reduced efficiency must be detected quickly and the treatment modified to lessen the effect. More analytical data must be collected and faster methods devised for using it. Because a wastewater treatment plant represents a large investment, efficient operation must be realized in order to obtain its benefits.

An important problem has been that of finding enough trained qualified personnel to operate municipal wastewater treatment plants. Unqualified personnel seem to have difficulty detecting and taking care of trouble with the result that an expensive treatment plant cannot be operated efficiently under its design criteria.[7]

The Cost of Textile Waste Treatment

There is a substantial amount of literature available on the problems of treating textile wastes but very little data on the cost of treating them. The information that is available is predominately from municipal facilities that may or may not be treating industrial wastes, especially those of textile plants, along with their domestic wastes.

Souther [8] estimated the costs of operation and maintenance (O and M) of the bioaeration plant at Cone Mills in Greensboro, N. C., to be about $0.07/1,000 gal (3,785 l) of wastes treated.

The four Cone Mills plants contributed 16.4 mil gal (61,994 cu m) of wastes for a total treatment cost of $1,154/wk in 1966. These four plants employed about 4,000 people for a 5.5-day (44-hr) week which, at $2/hr, would give a total payroll of $352,000/wk. The cost of waste treatment was only 0.3 percent of this total payroll.[9]

Geyer and Perry reported that costs for textile waste treatment varied from "10 cents to 20 cents/1,000 gal (3,785 l) for all items; labor, interest, depreciation, chemicals, and supervision."[10] "Water in Industry" reported the average waste treatment costs to be about $0.072/1,000 gal (3,785 l).[11] In general, waste treatment costs of municipal plants vary from $0.018 to $0.13/1,000 gal (3,785 l), depending on the size of the plant for complete treatment.[6]

Costs of waste treatment by means of chemical precipitation were reported to be higher than costs of conventional biological treatment. These costs varied from $0.055/1,000 gal (3,785 l) with a 20 percent biochemical oxygen demand (BOD) removal and 50 percent color removal to $0.57 with a 43.7 percent BOD removal and 97.5 percent color removal.[12]

Some plants have tried substituting low-BOD chemicals in order to reduce the pollution load. For example carboxymethyl cellulose (CMC), which has almost no BOD, has been used in place of starch for warp-sizing operations but at a slightly higher cost.[13] Because the bioadsorption process described by Souther[13] was a low-cost method for the disposal of starch from the desizing operation, this substitution may no longer be economical. Further results have shown the bioadsorption process to be one-third as costly as the conventional activated sludge or trickling filter treatments now in use by most municipalities.[12, 14]

As an alternative to other means of disposing of textile wastes, a lagoon or a mechanically aerated lagoon might prove satisfactory and economical.[11] Furthermore, a lagoon might serve a dual purpose because of the fact that the self-purifying capacities of streams are reduced during periods of low flow. Plant wastes could be stored in lagoons until stream flow returned to normal and then be released on a scheduled basis, thereby utilizing the assimilative capacity of the stream.

U. S. Public Health Service Publication Number 1229[16] presented construction cost data for most types of waste treatment plants. Rowan *et al.*[17] gave data on the O and M costs of various types of waste treatment plants. From the data in these two studies, which included data from municipal waste treatment facilities that handle industrial wastes, cost curves were derived by employing statistical techniques and were presented in cost-per-unit form for average flow, population (design or served) and population equivalents [one population equivalent equal 0.17 lb (0.771 kg) of BOD/day].

Moreover these two studies show that as plant size increases, the average per-unit cost decreases. Thus as plant size increases, the marginal (incremental) costs are lower than the average costs. As the size of a plant is increased, some costs remain constant over some given range of plant size. For example, if one doubles the size of a waste treatment plant, the construction costs increase, but only one plant operator may be needed. Doubling the size of a waste treatment plant may not always require doubling the number of inputs to operate the larger plant; consequently, economies of scale may exist.

Frankel[18] showed that for a given size of treatment plant the total annual cost per million gallons per day of wastes treated remained almost constant until one reached almost 80 percent BOD removal. As one moved above 80 percent BOD removal, the total annual costs per million gallons per day of wastes treated increased slightly until 90 percent BOD removal, and thereafter the annual costs increased sharply.

By use of reverse logic, one might conclude that once a treatment plant is designed to remove a certain percentage BOD, the total annual costs will remain constant whether it is removing BOD or not. Thus if one were to overload a waste treatment plant beyond its design capacity, costs would remain the same even though the efficiency of the plant would be reduced.

In another study, Nemerow [19] showed that textile wastes can be handled more economically in one municipal waste treatment plant than in two separate plants—one municipal and one industrial plant. Nemerow justifies this by assuming that the municipal plant can be designed to take care of most problems that would occur from treating textile wastes and that the rest could be taken care of by use of restrictions in the form of a municipal ordinance. He showed that there are many advantages such as only one plant operator, lower construction costs, and lower O and M costs. In addition, responsibility is placed on one specific owner, domestic wastewater adds nutrients and bacteria to organic industrial wastes, and a larger plant means more available funds for hiring better trained and qualified personnel. Also, a municipality is eligible for state and federal aid for construction of the treatment plant, whereas a privately owned textile plant is not eligible. For instance, in 1968, the federal government paid 30 percent of the costs if the state paid nothing, or up to 55 percent if the state paid 25 percent of the waste treatment plant costs, giving a maximum of 80 percent that the state and federal government combined might pay toward the costs of a municipal waste treatment plant.[20]

Conclusion

Because textile wastes at present have a very high BOD and are high in carbonaceous content, they should lend themselves to bacteriological decomposition as a method of treatment if the pH is not too high or low and if appropriate nutrients, which can be supplied by domestic wastes, are present. Thus it seems plausible to hypothesize that with current technology textile wastes can be treated economically in municipal treatment plants under certain restraints. But what may seem to be an adequate solution to the waste treatment problems of the textile industry today may be made obsolete tomorrow by advances in the textile process engineering or by new revolutionized treatment methods. Thus there is no simple solution to the problem of treating textile wastes and estimating their associated costs because of their changing nature from plant to plant and over a period of time.

Thus, from the review of literature, it seems that textile wastes can be treated economically in municipal wastewater treatment plants provided these wastes meet certain requirements that do not destroy the municipal plant efficiency.

Acknowledgment

The research for this work was supported in part by funds from the U. S. Department of the Interior, OWRR Project No. A-017-SC, J. M. Stepp, principal investigator, Clemson University Water Resources Institute.

References

1. Masselli, J. W., *et al.*, "The Effect of Industrial Wastes on Sewage Treatment." New Eng. Water Poll. Control Comm., Boston, Mass. **5**, 28 (June 1965).
2. Newlin, K. D., "An Economic Analysis of Treatment of Textile Wastes in Municipal Sewage Plants." M.S. thesis, Clemson Univ., S. C. (May 1969).

3. Denit, J. D., ''A Hydro-Economic Model of Industrial Waste Water Treatment Costs in South Carolina.'' M.S. thesis, Clemson Univ., S. C. (Aug. 1966).
4. U. S. Advisory Commission on Intergovernmental Relations, ''Intergovernmental Responsibilities for Water Supply and Sewage Disposal in Metropolitan Areas.'' U. S. Govt. Printing Office, Washington, D. C. (Oct. 1962).
5. Mantel, H. N., *et al.*, ''Industrial Incentives for Water Pollution Abatement.'' USPHS Report, U. S. Govt. Printing Office, Washington, D. C., 80 (Feb. 1965).
6. Imhoff, K., and Fair, G. M., ''Sewage Treatment.'' 2nd. Ed., John Wiley and Sons, New York, N. Y. 248 (1956).
7. ''Manpower and Training Needs in Water Pollution Control.'' FWPCA, U. S. Dept. of the Int., A report for 90th Congress, 1st Session, Senate Document No. 49, U. S. Govt. Printing Office, Washington, D. C. (Aug. 1967).
8. Souther, H. R., personal communication (1967).
9. Stack, V. T., Jr., *et al.*, ''Biological Treatment of Textile Wastes.'' *Proc. 16th Ind. Waste Conf.*, Purdue Univ., Ext. Ser. **109**, 387 (1962).
10. Geyer, J. C., and Perry, W. A., ''Textile Waste Treatment and Recovery.'' The Textile Foundation, Washington, D. C. 16 (1938).
11. ''Water in Industry.'' National Assn. of Mfg., New York, N. Y., 33, 46 (Jan. 1965).
12. ''Latest Word on Low-Cost Mill-Waste Disposal.'' *Textile World*, **110**, No. 6, 72 (1960).
13. Souther, R. H., ''Waste Water Control and Water Conservation.'' *Amer. Dyestuff Reporter*, **52**, 336 (1962).
14. Alspaugh, T. A., and Souther, R. H., ''Treating Textile Wastes Economically.'' *Textile World*, **108**, 336 (1958).
15. Kneese, A. V., ''The Economics of Regional Water Quality Management.'' The John Hopkins Press, Baltimore, Md. (1964).
16. ''Modern Sewage Treatment Plants—How Much Do They Cost?'' USPHS Publ. No. 1229, U. S. Govt. Printing Office, Washington, D. C., (1965).
17. Rowan, P. P., *et al.*, ''Estimating Sewage Treatment Plant Operation and Maintenance Costs.'' *Jour. Water Poll. Control Fed.*, **33**, 111 (1961).
18. Frankel, R. J., ''Economic Evaluation of Water Quality: An Engineering-Economic Model for Water Quality Management.'' San. Eng. Res. Lab. Rept. No. 65–3, Univ. of California, Berkeley, 10 (1965).
19. Nemerow, N. L., ''Theories and Practices of Industrial Waste Treatment.'' Addison-Wesley Publ. Co., Reading, Mass., 143 (1963).
20. ''The Economic Impact of the Capital Outlays Required to Obtain the Wastewater Quality Standards of the Federal Water Pollution Control Act.'' U. S. Dept. of the Int., FWPCA, U. S. Govt. Printing Office, Washington, D. C., 37 (Jan. 1968).

EFFLUENT QUALITY CONTROL AT A LARGE OIL REFINERY

Douglas S. Diehl, Robert T. Denbo, Manmohan N. Bhatla, and William D. Sitman

Humble Oil's Baton Rouge refinery was constructed on the Mississippi River in 1909 and has grown from an initial capacity of 2,000 bbl/day (318 cu m/day) to the present level of some 450,000 bbl/day (71,500 cu m/day). The refinery (Figure 1) manufactures a full line of petroleum products including motor and aviation gasolines; jet, diesel, and other distillate fuels; and lubricants and greases.

During the past 61 yr wastewater quality problems at the refinery have continually changed as industrialization and population along the river have increased. Early efforts in pollution abatement were directed primarily at improvements in treatment for reducing oil and phenol contents of effluent streams. During the 10-yr period ending in 1969, oil and phenol discharges have been reduced by 75 percent and 85 percent, respectively. This paper will report on the more recent developments and studies conducted for further effluent quality improvements.

The latest emphasis in wastewater quality improvement started in the early 1960's and culminated in a $19-mil project to switch from once-through river water for process cooling to a recirculating system with cooling towers. Once-through cooling has not been used for projects installed in the past 20 years and now accounts for 25 percent of cooling water capacity. The remaining 75 percent uses cooling towers. Once-through river water makes up 90 percent of the wastewater from the refinery. The emulsion of oil with silt from river water accounts for 80 percent of the oil discharged in the wastewater.

The river water replacement project was approximately 50 percent complete in October, 1970, and is scheduled for full completion by mid-1972. The exclusive purpose of this project is pollution abatement. An 80 percent reduction in oil contained in the effluent is expected along with a 90 percent reduction in the volume of wastewater. The reduction in volume will also permit additional treatment for removal of dissolved organics in equipment of reduced size.

The present in-plant refinery wastewater treatment facilities (Figs 2 and 3) include a sourwater stripper, a phenol extraction unit, ballast water treatment tanks, a silt deoiling plant, 14 oil-water separators, and a master separator, an earthen basin with a 2.5 hr detention time, for final oil removal. An 11-mil-gal (41,600-cu m) aerated lagoon (Figure 4) was recently started up to treat selected process streams.

Planning for improvement of wastewater quality in the refinery has been significant. Approximately 50 man-

FIGURE 1.—Humble Oil's Baton Rouge refinery.

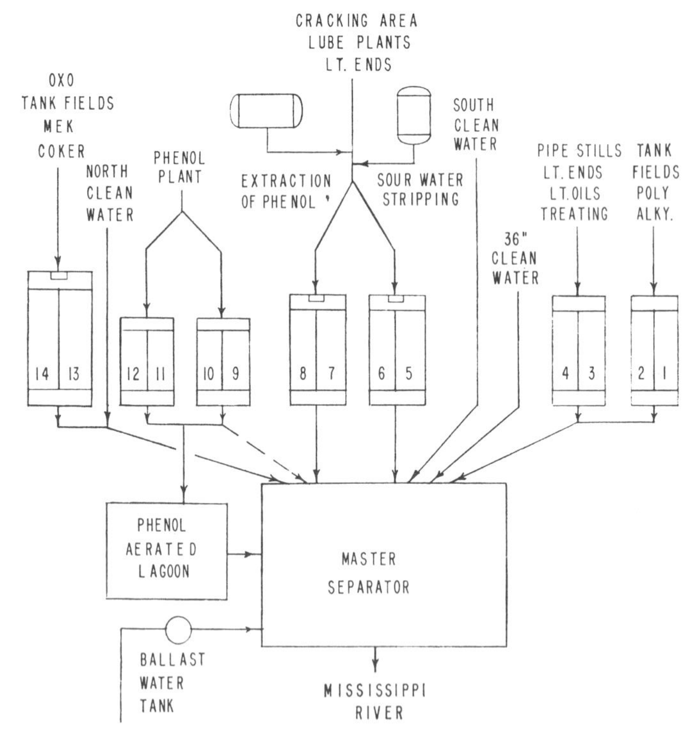

FIGURE 2.—Effluent treatment facilities, including oil-water separators and master separator. (In. × 2.54 = cm.)

FIGURE 3.—Existing wastewater treatment facilities.

years have been expended in these planning activities during the 1967–70 period.

During development of the river water replacement project in 1967, it was decided that a team approach to further wastewater quality improvement would be undertaken with the participation of a consulting firm. The effort to date on the studies described here have required approximately 25 man-years of effort with approximately 50 percent supplied by consultants' personnel.

After selection of a consulting firm, the studies were developed in six distinct phases:

1. Wastewater characterization;
2. Taste and odor studies;
3. Evaluation of joint and separate biological treatment of wastewater from the refinery and a nearby company-affiliated petrochemical plant:
4. Alternatives to biological treatment;
5. In-plant waste load reduction; and
6. Selection of additional treatment processes.

Wastewater Characterization

One of the first steps in the Baton Rouge refinery's current program was a complete characterization of its effluent and the in-plant definition of wastewater sources. This was accomplished primarily by an engineering consulting firm experienced in industrial wastewater treatment.

Over 325 individual waste streams were sampled and measured during the 6-month study. Analyses of biochemical oxygen demand (BOD), chemical oxygen demand (COD), phenol, oil, heavy metals, hydrogen sulfide, and solids were run as necessary—an average of three times on in-plant streams and ten or more times on the oil-water separator effluents. Flow measurements were often very difficult, and each determination presented a unique problem. Lithium dilution, Gurley meter, overflow weirs, or bucket and stop-watch were used to measure flow, depending on the situation. Flow measurement with a lithium chloride solution proved to be accurate and inexpensive in both sewers and pressure lines. A constant-head siphon was used to meter the solution into the sewers at a constant rate, and a chemical metering pump proved very successful for injection into pressurized lines.

The older parts of the refinery sewer system are very complex, and a large amount of time was needed to locate manholes and trace sewers with dye. Maps were updated and a numbering system for sewer sampling points was established. Material balances on flow and organic loading for each sewer system insured that all contributors to each sewer system had been correctly identified.

The survey detailed the character of the refinery's effluent and defined the statistical distribution of the more important parameters. The use of river water for once-through cooling results in a rather high rate of discharge to the Mississippi River. Elimination of this use (now under way) will reduce the total refinery effluent by almost 90 percent. This will put water usage below the average for refineries of similar complexity. Table I is a summary of wastewater quantities and characteristics at major units in the refinery, on the basis of units per barrel of refinery unit capacity.

The wastewater survey had many immediate as well as long-range effects. It increased the awareness of operating personnel of the importance of water pollution control. The study pointed out units that were major sources of wastewater contaminants and the units at which frequent upsets caused losses to sewers. Direct results of the study included recommendations to treat certain phenolic streams in an aerated lagoon in order to reduce the quantities of phenol being discharged to the river. This 4-acre, 11-mil gal (41,600-cu m) lagoon, equipped with three 75-hp (56-kw) mechanical aerators and having a detention time of 6 to 7 days (Figure 4) was started up in July, 1970. It has been successful in re-

TABLE I.—Wastewater Quantities and Characteristics from Refinery Units

Unit	Flow		BOD_5 (lb/bbl)	COD (lb/bbl)	Oil (lb/bbl)	Phenol (lb/bbl)
	Before RWRP* (gal/bbl)	After RWRP* (gal/bbl)				
Pipe still No. 7	16	8.5	0.020	0.045	0.53	0.0011
Pipe still No. 8	26	15.5	0.014	0.032	0.15	0.0005
Pipe still No. 9	11	6	0.022	0.061	0.10	0.0006
Pipe still No. 10	12	5	0.020	0.035	0.09	0.0006
Fluid catalytic cracker No. 2	10.4	10.4	0.035	0.071	0.181	0.0029
Fluid catalytic cracker No. 3	15.3	15.3	0.066	0.100	0.390	0.0079
Hydrocracker	29.2	29.2	0.200	0.304	1.210	0.0025
Powerformer	174	24.0	0.038	0.079	0.049	0.0011
Delayed coker	500	44.0	0.104	0.400	0.236	0.0157
Alkylation unit (sulfuric acid)	127	7.0	0.008	0.034	0.031	0.000
Propane lubrication plant	3,560	60	0.533	1.930	1.590	0.007
MEK Dewaxing	1,770	31	0.082	0.540	0.320	0.007

* River Water Replacement Project.
Note: gal/bbl × 0.0238 = cu m/cu m; lb/bbl × 3.1 × 10^{-3} = kg/kg.

FIGURE 4.—Aerated lagoon for treating odorous and phenolic wastes.

moving over 99 percent of the phenolics added to it and produces an effluent containing 0.005 to 0.05 mg/l of phenol. The study also illustrated the need for a detailed project to reduce further the amount and organic content of wastewater going to the sewers from each unit after completion of the river water replacement project. This study provided the information needed to select major waste streams for odor analysis.

Taste and Odor Study

The Mississippi River below Baton Rouge is the source of drinking water for about 1.5 mil people. Industrialization is heavy around the lower Mississippi (Figure 5), and during periods of low flow and low water temperature,

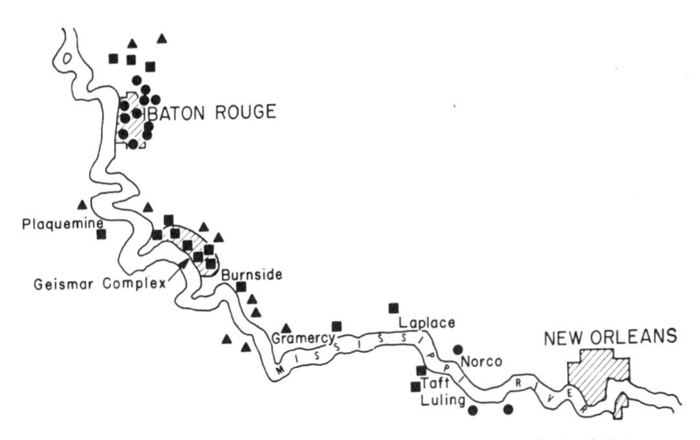

FIGURE 5.—Major industries along the lower Mississippi River.

some downstream water supplies have taste and odor problems.

In early 1969, the Federal Water Quality Administration (now the Office of Water Programs, Environmental Protection Agency), at the request of the Louisiana Stream Control Commission, established a field project office in Baton Rouge to study the taste and odor problems of the lower Mississippi. Although taste and odor had been a major concern of the refinery for many years, efforts were intensified concurrently with the establishment of this federal project. Humble organized a parallel study to define independently the refinery's taste and odor contribution to the river, to discover the major contributors of odorous wastewater in the refinery, and to study methods of reducing odor in various wastewater streams. The project was carried out by the consultant-refinery team and involved 2.5 man-years of effort over a 4-month period.

Wastewater streams were evaluated for odor by means of the threshold odor number (TON) as described in ASTM Standard Test D-1292-65. The TON is the greatest dilution of the sample with odor-free water, which can still be detected by an odor panel. For example, the odor in a wastewater with a TON of 8 can be detected when one volume of wastewater is diluted to 8 volumes with odor-free water. This test provides a fairly precise means for measuring apparent odor. Wastewater samples were exchanged between odor panels and the results were usually within 10 percent of each other.

Odor balances were performed in the refinery by the use of pseudo-quantitative "odor units," defined as TON × flow (gpm). In this manner, the team was able to perform "material balances" on sewer systems in terms of odorous materials to insure that all major streams had been sampled and to determine relative contributions of various streams.

The survey team began sampling at the refinery outfall and worked back into the plant. The eight major sewer systems discharging into the master separator were sampled to determine the relative importance of each system. From there, the major contributors to each sewer system were sampled. Only those wastewater sources suspected of being major odor contributors (based on known characteristics or field observation) were tested. Odor balances at various points in the system were used to insure that no major sources of odorous wastewater had been missed. Almost 100 wastewater streams were sampled in-plant from two to seven times each. Most of the more odorous streams discovered in the in-plant survey contained hydrogen sulfide. Subsequent investigation indicated that most of the hydrogen sulfide in these streams is either oxidized or stripped out in existing treatment facilities and the sewers before leaving the plant.

The survey showed that the majority of odor-causing compounds in the refinery wastewater are relatively unstable and are removed in the sewer system and separators by incidental stripping, volatilization, chemical oxidation, or reaction with other compounds. During periods of low river level when the refinery effluent falls 15 to 20 feet (4.6 to 6.1 m) over a spillway into the outfall canal, the remaining threshold odor is reduced an additional 40 percent. Coincidentally, it is during these periods of low river level that taste and odor are most critical downstream.

A portion of the odor-causing compounds were found to be readily biodegradable but not easily stripped or chemically oxidized. Another small persistent portion seemed to be resistant to biological degradation. An odor persistency test has recently been developed at the refinery to differentiate between these persistent and nonpersistent odors in the various refinery waste streams. This qualitative test

TABLE II.—Sources of Odorous Wastewater

Refinery Unit	Wastewater Source	TON
Pipe stills	Barometric condensers*	4,600–31,000
Catalytic cracking units	Fractionator overhead accumulator drums†	135,000
Light oils finishing unit	Sweetening unit water wash†	170,000
Emulsion breaking tanks	Water draws†	3,000–7,000
Gas flares	Seal drums‡	0–150,000
Light slop spheres	Water draws§	150,000
Pipe still	Overhead accumulator drums	1,000–3,400
	Desalter brine	200–600
Delayed coker	Total unit	1,000–3,000
Hydrogen sulfide stripper	Sour water	3,000

* Scheduled for replacement with surface condensers.
† Scheduled for treatment in aerated lagoon.
‡ TON is high only when flare is burning.
§ Primarily H_2S and NH_4.

consists of diluting the wastewater to a preset total organic carbon (TOC) or COD concentration with odor-free water, adding a small amount of acclimated biological seed, and aerating it gently at room temperature. TON analyses are performed initially and then daily for 4 days. The comparative slope and residual odor are used as an indication of the persistency of the wastewater odor.

The in-plant sampling indicated that the crude distillation equipment accounts for the majority of the odor in the refinery effluent. The barometric condensers on the vacuum distillation towers at the pipe stills accounted for almost 50 percent of the odor in the refinery effluent. These barometric condensers use once-through river water to condense steam that contains some products of cracking from the vacuum furnaces. Typical TON's from these barometrics range from 4,600 to 31,000 with flows from 340 to 1,100 gpm (22 to 70 l/sec) each. Fortunately, these barometric condensers are being replaced by surface condensers cooled by recirculated cooling water as part of the river water replacement project (RWRP). As a result of this taste and odor survey, the installation of these surface condensers has been expedited. These process changes and the increased detention times in the wastewater collection and treatment system resulting from decreased flow should result in a 50 percent or more reduction in the odor contribution to the river compared to 1969 levels. Table II is a list of the more odorous waste streams found in the refinery and their threshold odor numbers.

While the odor survey was being completed, laboratory treatability studies were undertaken to determine methods by which wastewater odor could be reduced. The studies included biological treatment, air stripping, plain detention, activated carbon adsorption, and ozonation.

Aerated Lagoon

A bench-scale aerated lagoon was operated in the laboratory for approximately 5 wk to determine its ability to treat seven of the more odorous and/or phenolic waste streams. The unit performed well, with over 90 percent odor reduction, 99 percent phenol reduction, and 97 percent BOD reduction observed with a 5.5-day detention time. Based on these tests, the decision was made to add four additional 75-hp (56-kw) floating aerators to the existing lagoon. When these aerators are installed in late 1970, the lagoon is expected to reduce the threshold odor leaving the refinery by an additional

10 to 15 percent while destroying a significant amount of phenolics.

Activated Sludge

To define the amount of refinery wastewater odor reduction obtainable by the activated sludge process, a continuous-flow bench-scale unit was operated for approximately 11 wk. The unit consisted of a 15-l aeration tank and 7-l conical settling basin equipped with variable-speed feed and sludge return pumps.

Because the characteristics of the refinery wastewater after the completion of the RWRP will be quite different, two types of wastewater were fed to the activated sludge unit. For the first 5 wk, present refinery effluent was used as the feed. For the remaining 5 wk, a composite of four preselected streams simulating the post-RWRP wastewater was fed to the unit. These two feed stocks were quite different in character and strength, and the results obtained from operating the unit were also different.

Table III summarizes the results obtained from the two feed stocks. Activated sludge with 2 hr aeration time was capable of achieving greater than 80 percent odor reduction from the present refinery effluent. When operated on the stronger composite of in-plant waste streams, the unit was able to achieve slightly better than 65 percent reduction in threshold odor with a detention time of 5 hr. Apparently, the more limited number of compounds in the preselected in-plant streams were more difficult to degrade than the broader range of odor-causing compounds in the present refinery effluent.

Tests performed concurrently with these studies indicated that the air stripping and mixing in the shallow 15-l aeration tank accounted for 20 to 35 percent of the odor reduction achieved in each case. The remainder of the odor reduction is therefore attributed to biological degradation.

TABLE III.—Bench-Scale Activated Sludge Performance

	Feed	Effluent	Reduction (%)
Refinery effluent*			
Odor (TON)	268	40	83
COD (mg/l)	152	63	55
BOD_5 (mg/l)	28	5	86
Composite of in-plant streams†			
Odor, TON	2,380	690	66
COD (mg/l)	1,290	680	63
BOD_5 (mg/l)	450	40	92

* Two-hour aeration time with mechanical mixing and a small amount of diffused air.
† Five-hour aeration time with mechanical mixing and a small amount of diffused air. Waste composition: flare seal drum water when burning, 6%; olefins unit distillate drum, 18%; combined wastewaters from No. 1 chemical unit, 38%; pipe still No. 10 barometric condenser water, 38%.

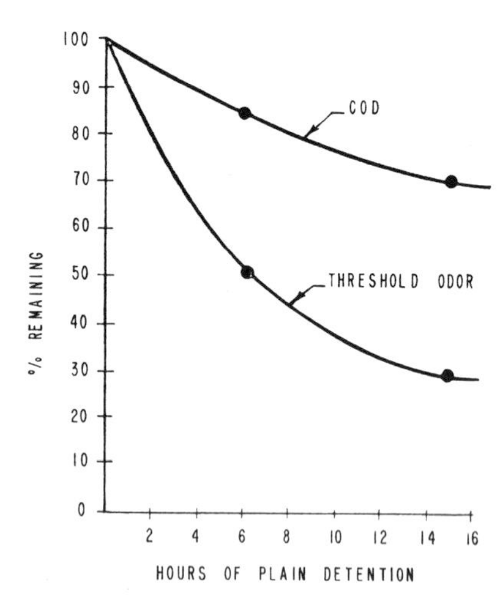

FIGURE 6.—Qualitative effect of detention on refinery effluent.

Air Stripping

During the course of the biological odor reduction investigations, it became apparent that aeration and plain detention were responsible for significant decreases in the wastewater threshold odor and, to a lesser degree, COD. Additional tests on the present refinery effluent using an 8-ft (2.44-m) column indicated that odor reductions of 50 percent or more can be obtained by aerating with 2 to 4 volumes of air/volume of wastewater in deep tanks with detention times of 10 to 20 min.

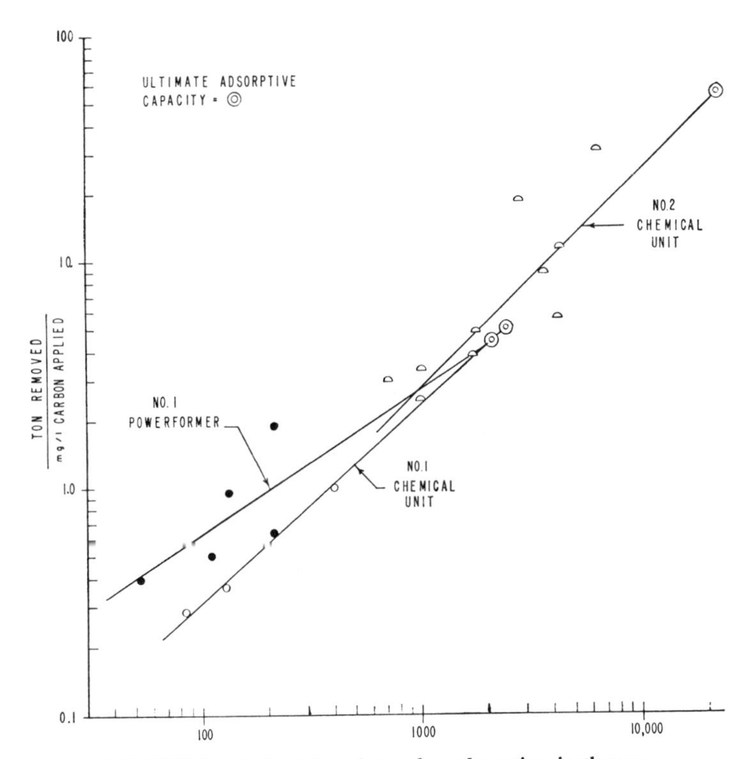

FIGURE 7.—Activated carbon odor adsorption isotherms.

Plain Detention

Tests were also run to determine the effects of holding samples of the present refinery effluent. As shown in Figure 6, studies indicated that with 6 hr of plain detention, a 50 percent TON reduction and 15 percent COD reduction occurred. With 15 hr detention time, a 70 percent TON reduction and 30 percent COD reduction were observed. The reasons for the observed reductions in odor and COD are not known precisely. However, chemical oxidation, volatilization, and oil separation are thought to contribute to the reduction, and it appears that plain detention of this refinery wastewater, which can be provided by equalization basins, aerated lagoons, clarifiers, and so on, will result in significant reductions in wastewater odor.

Activated Carbon

Carbon adsorption isotherms were run to determine the amenability of the odor-causing compounds found in this wastewater to removal by activated carbon. The isotherms were determined for several wastewater streams whose relatively high threshold odor levels were caused by compounds other than hydrogen sulfide.

Six odor adsorption isotherms are shown in Figures 7 and 8. The wastewater samples were collected immediately before the test, settled for 30 min, and filtered under a low vacuum to remove any remaining oil. Varying amounts of powdered activated carbon were added to each flask except the one used as the control. The flasks were then agitated uniformly for 60 min in a constant-temperature bath.

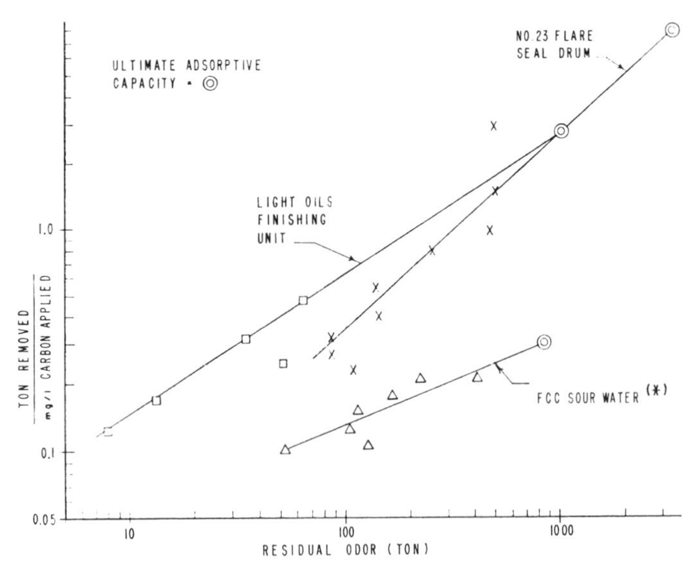

FIGURE 8.—Activated carbon odor adsorption isotherms. (*Air-stripped in laboratory.)

At the end of the agitation period, the samples were filtered under a low vacuum to remove the carbon and analyzed for odor and COD. The control sample was used as the base line, as the agitation and filtering accounted for a portion of the odor reduction observed.

The odor isotherms shown in Figures 7 and 8 can be used to predict the amount of powdered carbon required to reach a certain threshold odor level for wastewater. Additionally, by extrapolating the line drawn through the data points back to the TON of the control wastewater, the theoretical adsorptive capacity of granular carbon in fixed beds can be determined. This was done for each of the six isotherms presented, and the results were plotted against wastewater TON as shown in Figure 9. As expected, the adsorptive capacity increased as the wastewater TON increased. The only exception was the unstripped cat cracker sour water which was air-stripped to remove hydrogen sulfide. Many of the more easily adsorbed volatile odor-causing compounds were apparently stripped off.

These adsorption isotherms demonstrated, on a preliminary basis, that activated carbon is effective in removing the odor-causing compounds in this

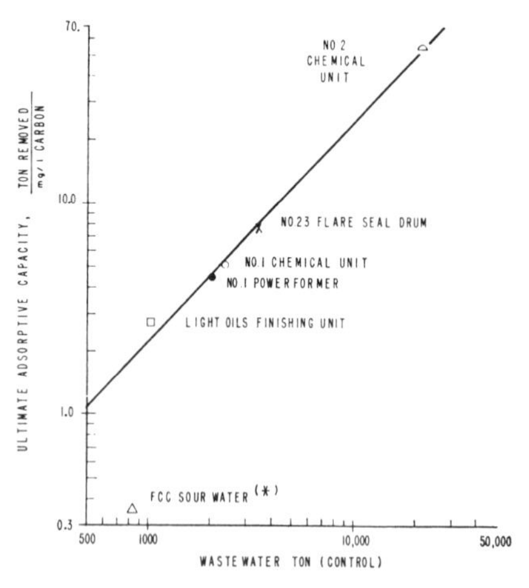

FIGURE 9.—Activated carbon odor adsorptive capacity. (*Air-stripped in laboratory.)

TABLE IV.—Costs of Biological Sludge Dewatering Unit Processes*

Process	Solids In (%)	Solids Out (%)	Capital Cost	$/Ton†
Aerobic digestion	—	—	$ 190,000	2.40
Centrifugation				
Disks	3–4	5–6	212,000	2.60
Disks and solid bowls	3–4	15	530,000	6.25
Vacuum filtration	3–4	15	489,000	12.10
Filter Presses				
Manufacturer A	3–4	45	1,450,000	18.90
Manufacturer B	3–4	45	1,383,000	18.40
Heat treatment				
Manufacturer C	4	35	2,550,000	21.00
Manufacturer D	6	35	2,330,000	18.00
Flash drying (no incineration)	30‡	80–85	1,136,000	14.10
Low pressure				
wet oxidation	4	—	2,570,000	19.65
	6	—	2,240,000	17.00

* 75 tons/day (68 metric tons/day) dry solids.
† Includes depreciation, interest, insurance, operation, and maintenance.
‡ Following filter presses.
Note: $/ton × 0.907 = $/metric ton

refinery wastewater. The costs involved and other questions will be discussed later.

Ozonation

During the project, the use of ozone was evaluated on three wastewaters. The tests indicated that ozone was generally responsible for some reduction in COD and BOD. However, ozonation was found to have an unusual effect on the threshold odor of the wastewater streams tested. Whereas plain oxygenation of the control wastewater caused the threshold odor to decrease continually, ozonation caused the threshold odor to decrease to a certain level and then to increase. In addition, the character of the odor changed from a hydrocarbon odor to a sweet-sour odor similar to acetic acid.

Joint and Separate Biological Treatment

Humble and a neighboring affiliated petrochemical plant are studying joint and separate biological treatment because biological treatment is economical for dilute organic wastes. These studies considered the use of a number of separate sites for both aerated lagoon and activated sludge treatment of Humble's refinery wastewater alone or for the combined refinery and petrochemical plant wastewater. Additionally, the treatment of the combined wastewater with and without pretreatment of the chemical plant wastewater by means of a roughing filter was considered.

Cost estimates were prepared for each of the sites and treatment schemes outlined above. The study has determined that short-range treatment requirements with emphasis on odor reduction can be achieved by either the aerated lagoon or activated sludge process at sites considered, for either the combined refinery-chemical plant waste streams or for the refinery alone. Any of the sites considered for an activated sludge plant would be very costly. Moreover, this route may not completely solve the refinery's wastewater quality problems. Consequently, a definitive treatment scheme has not yet been selected.

Biological Sludge Disposal

A separate detailed study was conducted to evaluate the costs of various methods of biological sludge disposal at the Baton Rouge refinery. Over 25 dewatering and disposal processes were studied for three different rates of sludge production—75, 150, and 250 tons (67, 136, and 227 metric tons) of dry solids/day. Tables IV and V summarize the capital and unit costs for the dewatering unit processes and disposal methods. The information provided by this sludge disposal study has provided valuable input to the activated sludge cases studied to date and will be most useful in making future decisions.

Alternatives to Biological Treatment

An activated sludge facility with the necessary equalization and sludge disposal facilities would require about 15 to 20 acres (6.1 to 8.1 ha) of land for wastewater from the refinery and petrochemical plant. The refinery would prefer to avoid the problems inherent in the operation of a biological system on refinery wastewater such as upsets resulting from toxic discharges or organic surges and the problems of sludge handling and disposal. It is also possible that biological treatment may not completely solve the taste and odor problems. For these reasons and because of high land requirements for biological processes, alternate treating schemes were considered. These included activated carbon, coagulation-sedimentation, and dissolved air flotation. Laboratory tests of these processes were performed using various strong wastewaters which will still be present after completion of the River Water Replacement Project. In addition, reverse osmosis was studied by reviewing current literature and holding discussions with various equipment manufacturers.

Activated Carbon Adsorption

As described previously, activated carbon adsorption isotherms were run on several strong wastewater streams. The wastewater samples were analyzed for both odor and COD. Isotherms similar to the ones for odor were plotted for COD (Figure 10). Again, these isotherms were extended back to the COD concentrations of the control wastewater in order to estimate the theoretical adsorptive capacity of granular carbon for each wastewater when used in a fixed bed. The theoretical adsorptive capacity (Figure 11) varied widely from wastewater to wastewater and showed no relation to the original COD concentration. This is attributed to the fact that the types of organic compounds in the wastewater varied widely from stream to stream. The carbon could more easily adsorb some organic com-

TABLE V.—Costs of Biological Sludge Disposal*

Manufacturer	Solids (%)	Capital Cost	$/Ton†
Fluid bed incineration			
Manufacturer A	15	$1,035,000	23.80
Manufacturer B	15	4,680,000	54.15
Multiple hearth incineration			
Manufacturer C	15	2,340,000	29.80
Manufacturer D	15	1,640,000	24.95
Manufacturer E	15	865,000	16.40
Contract incineration	10	—	65.95
Company A	15	—	43.90
Contract barging to gulf†	10	—	16.00–33.40
	15	—	13.30–22.30

* 75 tons/day (68 metric tons/day) dry solids.
† Includes depreciation, interest, insurance, operation, and maintenance.
Note: $/ton × 0.907 = $/metric ton.

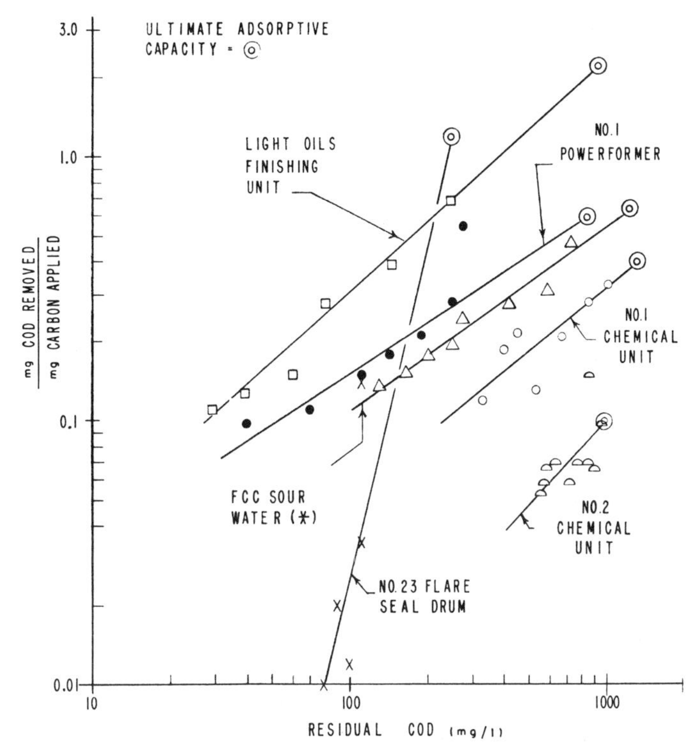

FIGURE 10.—Activated carbon adsorption isotherms. (*Air-stripped in laboratory.)

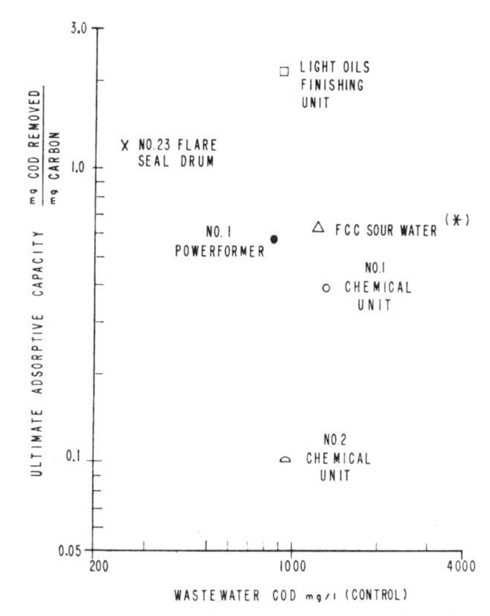

FIGURE 11.—Activated carbon chemical oxygen demand adsorptive capacity. (*Air-stripped in laboratory.)

pounds than it could others. However, the odor-causing compounds in the various waste streams all exhibited a similar degree of adsorption. This is attributed to the fact that the majority of the odor-causing compounds are probably moderately volatile and apparently of a similar intermediate molecular weight.

To help determine on a preliminary basis the potential of activated carbon for treatment of Humble's wastewater, the isotherm data obtained during this study were used to estimate order-of-magnitude carbon requirements and treatment costs. Two design cases (both after river water replacement, but before in-plant modifications) were considered: (*a*) a system designed for 80 percent COD removal (corresponding to 85 to 90 percent BOD reduction), and (*b*) a system designed for

90 percent odor reduction of Humble's wastewater. For purposes of comparing investment, cases were prepared for a refinery wastewater (after the oil-water separators) containing 670 mg/l of COD, the equivalent of 121,000 lb/day (55,000 kg/day), with a TON of 5,000 at a flow rate of 22 mgd (83,300 cu m/day). Pretreatment such as dissolved air flotation and/or filtration was assumed to remove a total of 20 percent of the COD and threshold odor before carbon treatment, and these removals were given credit in each of the cases.

The carbon regeneration requirements of 290,000 lb/day and 140,000 lb/day (131,500 and 63,500 kg/day) for Cases 1 and 2, respectively, were based on the assumption that a granular carbon system could achieve 50 percent of the ultimate adsorptive capacities indicated by the powdered carbon isotherms. This is a conservative approach that neglects any benefits of biological activity but takes into account the reduced adsorptive capacity of the carbon after several regeneration cycles. The COD adsorptive capacity of 0.50 lb COD/lb of powdered carbon is an average of values obtained during these isotherm tests and is typical of values obtained from other refinery wastewater. The odor adsorptive capacity of 9.5 TON removed per mg/l of powdered carbon applied was obtained from Figure 9, for refinery wastewater with a TON of 4,000 (that is, after dissolved air flotation). Therefore, the granular carbon regeneration requirements listed above are based on adsorptive capacities of 0.25 lb of COD/lb of carbon and 4.75 TON removed/mg/l of carbon.

Total costs were calculated by estimating the total cost of carbon regeneration at $0.02/lb ($0.044/kg) and by estimating the remaining total costs at approximately $0.10/1,000 gal ($0.026/cu m), including interest charges on capital investment. These

TABLE VI.—Activated Carbon Treatment of Refinery Wastewater*

	Case 1	Case 2
COD Removal (%)†	80	50
Odor Reduction (%)†	95+	90
Estimated cost/1,000 gal‡	$0.37	$0.23
Possible range of costs/1,000 gal‡	$0.13–$0.66	$0.13–$0.43

* Oil-water separator effluent COD = 670 mg/l, TON = 5,000, and flow = 22 mgd.

† Includes 20 percent removal by dissolved air flotation.

‡ Excludes cost of pretreatment. Includes operation, maintenance, and capital amortization.

Note: $/1,000 gal × 0.264 = $/cu m.

values are based on discussions with an activated carbon manufacturer, values from the literature, and past experience with activated carbon systems.

Because odor-causing compounds are preferentially adsorbed by activated carbon, the cost of removing 80 percent of the COD is greater than the cost of removing 90 percent of the odor.

For a system designed to remove 80 percent of the COD (Table VI), the most probable cost is estimated at $0.37/1,000 gal ($0.098/cu m), with a possible range of $0.13 to $0.66/1,000 gal ($0.034 to $0.174/cu m). For a system designed and operated to remove 90 percent of the odor, the most probable cost is estimated at $0.23/1,000 gal ($0.061/cu m), with a possible range of $0.13 to $0.43/1,000 gal ($0.034 to $0.114/cu m). This latter system would also remove about 50 percent of the COD including that accomplished by pretreatment. The on-site capital costs of an activated carbon system, including regeneration facilities, are estimated at $3.5 mil and $3.0 mil, respectively, for each of the above cases, but could range between $2 and $4 mil. None of the above costs includes the cost of any pretreatment because the degree of pretreatment necessary depends on many factors not yet developed. The ranges of costs given above reflect the most favorable and unfavorable values expected, based on wastewater composition, future technological developments in carbon application techniques, handling, and regenerating, and the effect of biological activity in the carbon columns.

Several questions remain to be answered before the complete feasibility of treating Humble's wastewater with carbon can be determined. These include:

1. Practical degree of organics removal obtainable by carbon alone on some types of streams;
2. Role and optimization of biological activity in carbon beds;
3. Optimum design of carbon columns;
4. Pretreatment requirements, as related to Items 2 and 3;
5. Backwash requirements as related to Items 2, 3, and 4;
6. Adsorptive capacities actually achievable in a fixed bed;
7. Effect of the concentration of sodium and other salts on the regeneration of carbon.

Because the character of the refinery effluent will change, the final activated carbon column studies will not be undertaken until after the River Water Replacement Project is complete in mid-1972. In the meantime, intermediate studies are planned.

Coagulation-Sedimentation and Dissolved Air Flotation

Coagulation-sedimentation and dissolved air flotation studies were undertaken to determine applicability of these unit processes. Laboratory tests on coagulation, settling, and dissolved air flotation were run on a series of wastewaters representing streams that would remain after completion of the RWRP.

Coagulation-sedimentation using 25 to 95 mg/l of ferric chloride or 30 to 80 mg/l of alum [as $AL_2(SO_4)_3$] was generally successful in reducing COD and oil concentrations in the wastewater. COD reduction averaged 50 percent, ranging between 0 and 90 percent, and oil reductions averaged almost 60 percent, ranging between 20 and 99 percent. Dissolved air flotation achieved similar degrees of COD and oil removal, although smaller dosages of coagulants were usually required. Dissolved air flotation without chemical coagulation generally did not significantly reduce either COD or oil concentration. Removal by coagulation-sedimentation or dissolved air flotation, as well as the optimum coagulant dosages, varied widely from wastewater to wastewater, illustrating the need for further consideration after completion of the RWRP.

Reverse Osmosis

After the literature on reverse osmosis (RO) was reviewed and several manufacturers of RO equipment were consulted, the conclusion was reached that at the present level of development, RO is not economically feasible for wastewater treatment at the Baton Rouge refinery.

High-Rate Filtration

Preliminary screening of high-rate filtration for oil and suspended solids removal has been conducted. Future work is planned with selected streams to determine potential refinery applications of this process.

In-Plant Wastewater Reduction

Before extensive wastewater treatment facilities are built, it is logical to reduce the amount and character of the wastewater by process and equipment modifications, water reuse, product recovery, and operational changes. The 18-month construction period for the RWRP, which will reduce wastewater volume by 90 percent, provided an ideal period for an intensive inplant wastewater reduction program. This type of program is the most positive method of reducing pollution be-

TABLE VII.—Estimated Reductions in Wastewater by In-Plant Modifications

Unit	Flow			BOD_5			Threshold Odor		
	1969 (gpm)	After River Water Replacement (gpm)	After In-Plant Reduction (gpm)	1969 (lb/day)	After River Water Replacement (lb/day)	After In-Plant Reduction (lb/day)	1969 (odor units*)	After River Water Replacement (odor units*)	After In-Plant Reduction (odor units*)
Pipe still No. 7	820	450	185	1,500	900	800	5.3×10^6	300,000	200,000
Pipe still No. 8	1,170	700	130	900	600	500	10.8×10^6	400,000	300,000
Propane lube plant	17,300	290	110	3,700	700	650	3.1×10^6	700,000	500,000
MEK plant	16,000	280	55	1,100	600	500	—	—	—
Phenol treating plant	425	425	205	110	110	100	1.0×10^6	1.0×10^6	900,000
No. 3 light ends unit —north	4,000	450	35	600	50	50	400,000	100,000	90,000

* Odor unit = TON × flow (gpm).
Note: gpm × 0.0631 = 1/sec; lb/day × 0.454 = kg/day.

cause 100 percent of the pollutants kept out of the system are removed rather than the 60 to 85 percent removal in most treatment processes.

The in-plant wastewater reduction project was scheduled so that the modifications could be completed as the various units were converted from river water to recirculated water. Humble began the wastewater reduction project in June 1970, and the first new cooling towers were scheduled for start-up in December 1970. Each of the 33 refinery units are being studied by one of two teams, each consisting of one refinery engineer, one consulting engineer, one refinery engineering technican, and an engineering draftsman. At each unit, the process and sources of wastewater are being studied with the help of the unit supervisor and technical contact personnel. To supplement existing information, flow measurements are made and samples collected for analysis. Meetings are held with technical and operating personnel to discuss potential means of reducing wastewater quantities and pollutants. A report on each unit lists all wastewater sources and describes all proposed modifications. A simplified diagram, preliminary cost estimate, responsibilities for accomplishment, and desired completion date are prepared for each recommendation. From this report, mechanical work orders are issued for the small modifications, and process specifications with definitive designs are prepared for the larger modifications. Priorities are being developed for the installation of the facilities defined.

One of the important aspects of this project is the pollution awareness training given to the refinery operating personnel. As each team visits a unit, a short talk is given to the operating personnel outlining the refinery's commitment to water quality improvement and telling them how they personally can help with the program. These personnel are then involved with the team members during the survey, and, by means of this involvement, their awareness of in-plant abatement is increased.

As of October 1970, the two teams had covered over half of the units in the refinery, and modifications had been started at some units. The results attained at the first units were more significant than expected. Table VIII lists the wastewater reductions expected at the first units surveyed as a result of the RWRP and the wastewater reduction project. Wastewater reductions of similar magnitude are expected at the remaining units.

In addition to the unit-by-unit study, refinery-wide projects are being undertaken by the teams. These include detailed examination of the refinery slop oil system and procedures and chemicals used for chemical cleaning.

Selection of Additional Treatment Processes

Following completion of the RWRP in mid-1972, final treatability studies will be undertaken to gather design data for both biological and physicochemical treatment processes. Activated sludge, activated carbon adsorption, dissolved air flotation, and high-rate filtration will be studied in depth. Once the necessary data are collected and developed, the various treatment alternatives will be studied using the systems approach to select optimum treatment needed.

Summary

In summary, this paper has attempted to show how a large complex oil refinery has undertaken a comprehensive program to upgrade the quality of its wastewater with the help of an environmental engineering consulting firm. A combination of personnel with varied knowledge and expertise has been used to define the critical effluent parameters, investigate treatment processes, reduce water pollution at the source, and apply a systems approach to the design of effective treatment facilities.

Projects recently completed or under construction will reduce the refinery's effluent flow by 90 percent, oil load by 80 percent, odor units by 70 percent, BOD load by 30 percent, and phenol load by 40 percent. This does not include the significant improvements expected from the in-plant wastewater reduction project that is currently in progress.

Acknowledgments

The authors wish to express their appreciation for the assistance and contributions of N. D. Radford, W. F. Milbury, A. F. Thompson, K. G. Barnhill, and A. B. Chandler of Roy F. Weston, Inc. and C. L. McVea, F. W. Gowdy, W. B. Ogles, M. R. Aubin, and G. F. Ullrich of Humble Oil & Refining Co.

EVALUATION CRITERIA FOR GRANULAR ACTIVATED CARBONS

James S. Mattson and Frank W. Kennedy

The growing use of granular activated carbon in tertiary and physical-chemical wastewater treatment today,[1-14] combined with the high cost of carbon [$0.26 to $0.29/lb ($0.57 to $0.64/kg)], requires the development of improved analytical methods for technical evaluation of granular activated carbons. As a general rule, suppliers of activated carbon provide a table of quality parameters for their products comparable to Table I. Most of this information is of little or no value to the individual whose responsibilities include evaluating the economics of several competing activated carbons. Granted that the N_2-BET surface areas, iodine numbers, and molasses numbers are related to the ability of activated carbons to adsorb dissolved organics from wastewater, these numbers nevertheless tend more to confuse the issue than to clarify it. The abrasion number and "regeneration loss" values reported can rarely be interpreted to have any quantitative meaning for the application at hand. Actually, it would be better if potential users of granular carbon for wastewater treatment undertook an evaluation of each manufacturer's product in their own systems and ignored the data provided by the suppliers. In this laboratory, it has often been observed that the products received from the various manufacturers do not meet the specifications listed in the accompanying data sheets.

In the general area of wastewater treatment by granular activated carbon adsorption, the economics are such that the carbon columns are utilized best by driving them to their equilibrium adsorption capacity and then regenerating and reusing the carbon as many times as possible. The number of regeneration cycles obtained in field operations at tertiary treatment pilot plants[12-14] has been from 10 to 20 cycles. The cost of treating wastewater with carbon that yields a 10 percent regeneration loss is obviously greater than that for a carbon that undergoes only a 2 percent regeneration loss.[5] Also, an increase in the adsorption capacity for any carbon can be economically offset by a corresponding decrease in its ability to resist attrition.

The third parameter of importance in this area is the rate of adsorption by granular activated carbon. The rate of adsorption determines the contact time and flow rate required to produce a given quality of effluent. This obviously influences the capital cost requirements for a treatment plant and, because of the high cost of obtaining such capital today, should play a major role in determining the supplier of granular carbon.

These three parameters, adsorption capacity, rate of adsorption, and resistance to attrition, are examined in turn in the following sections. Some representative data obtained for commerical and laboratory-produced granular activated carbons are presented as examples.

Adsorption Capacity

Adsorption of dissolved organic weak electrolytes and nonelectrolytes by active carbon is reviewed elsewhere,[15] and the nature of the dissolved organic species found in municipal wastewaters has received some attention in the literature.[16, 17] Unfortunately, the characteristics of wastewaters vary significantly in time and space. Therefore, no adequate method exists for examining the capacity of a carbon to adsorb the dissolved organic components to be found in wastewater per se. However, activated carbons have been used extensively for the adsorption and recovery of phenols from municipal water supplies and industrial effluents. The adsorption of phenol and substituted phenols has been thoroughly discussed in the literature.[15, 18-24] The differences observed among various types of activated carbons for adsorption of substituted phenols and the relative ease of obtaining the adsorption isotherms for these species lend credence to the use of such measurements as characteristic of the adsorption capacity of activated carbons.

One manufacturer's literature[25] acknowledges that the notion of the N_2-BET surface area really has little relevance to the application of activated carbon in wastewater treatment. In fact, capacities of only about 15 percent of the N_2-BET surface area[22] are typical of those observed for *p*-nitrophenol (PNP) adsorption. Phenol itself, which exhibits a two-step isotherm on activated carbon,[26] covers as much as 45 percent of the total N_2-BET surface area, but only at very high concentrations. Phenol is not a satisfactory solute for adsorption isotherms for several reasons: its high solubility in water, the difficulty with which that solubility is measured, and its tendency to form water-in-phenol colloidal suspensions. However, PNP has a lower water solubility, is fairly easy to purify, and is easily analyzed quantitatively by ultraviolet spectrophotometry. Also, because of its low solubility in water, PNP is adsorbed to a much greater extent than is phenol at micromolar concentrations.[22]

TABLE I.—Ranges of Evaluation Parameters Given for Commercial Granular Activated Carbons

Parameter	Units	Range
N_2-BET surface area	sq m/g	700–1,100
Pore volume	cu cm/g	0.60–1.0
Molasses decolorizing index	—	6.3–8.0
Iodine number	—	900–1,100
Apparent density	lb/cu ft	24.5–30
Water-wetted density	g/cu cm	1.3–1.4
Abrasion number	—	70–80
Ash content	%	0.5–8.5
Moisture content	%	2–12
pH of aqueous extract	—	6–8
Water solubles	%	1
Regeneration loss	%	5

The analytical procedure for obtaining PNP isotherms is as follows:

1. A 10^{-3} *M* solution of purified PNP (recrystallized from toluene until it has a melting point of 113° to 114°C), is prepared at a pH of 2 in distilled water. Ionic strength is an important factor, so the distilled water should have a specific conductivity below 1 to 2 μmhos/cm. The pH of 2 is obtained with HCl. The solubility of PNP is strongly pH dependent, as is the isotherm, and the isotherm is slightly ionic-strength dependent.

2. The test carbon is pulverized to −325 (U.S. Sieve) mesh in order to facilitate determination of the adsorption isotherm within a reasonable length of time. Inorganic ash can be removed by washing with 6 *M* HCl and rinsing several times in distilled water. Fines are removed by allowing the larger particles to settle in the rinse water for about an hour and decanting. The carbon is dried for 24 hr at 130°C and stored in a vacuum desiccator until used.

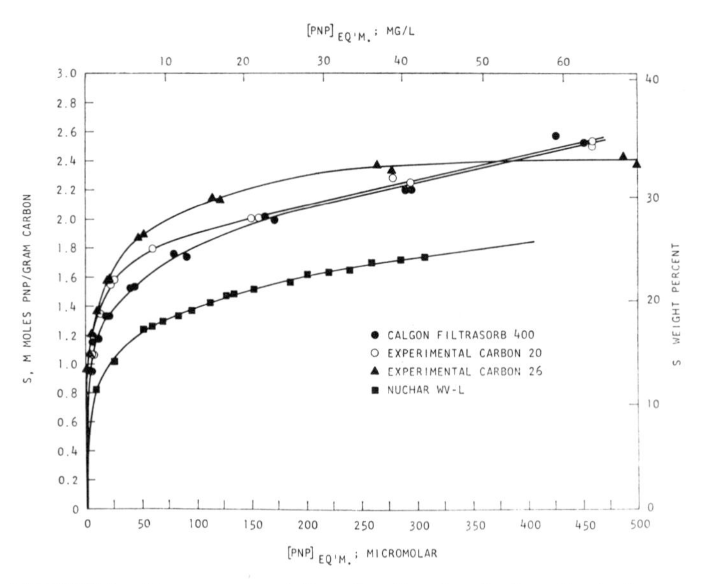

FIGURE 1.—Linear representation of *p*-nitrophenol adsorption isotherms of two commercial and two experimental activated carbons.

3. The "bottle-point" method of determining the isotherm is used. Twenty clean French-square 4-oz (0.12-l) glass bottles with polyethylene-lined screw caps are washed with distilled water and dried before use. Duplicate samples of 20, 30, 40, 50, 60, 70, 80, 90, and 100 mg ($\pm$10 percent) of the sample carbon are weighed into 18 of the dried bottles. Next, 100 ml of the 10^{-3} *M* PNP (pH 2) is pipetted into each of the 20 bottles (2 of which serve as blanks). The bottles are capped, and the samples are equilibrated for a minimum of 2 hr and no more than 24 hr. The samples are kept in continuous agitation. This is accomplished by placing them in a slowly rotating device that turns them end-over-end at about 6 rpm. After the equilibration period is over, the samples are filtered through 0.45-μ microporous filters using a clean filtering apparatus, and the filtrate is collected for analysis. The apparatus can be washed between samples with methanol. Acetone should be avoided because it interferes with the analysis. The filtrate samples are then analyzed colorimetrically at 317 mμ (molar absorption coefficient = 9,800), using pH 2 water for dilution. The amount of PNP adsorbed is then calculated, with the average PNP concentration of the blanks used as the starting concentration. The amount adsorbed (in m*M* PNP/g carbon) is plotted versus the equilibrium PNP concentration.

Figure 1 illustrates the PNP adsorption isotherms obtained for four different activated carbons. It can be seen that the adsorption capacities *S* of the four activated carbons are significantly different. The adsorption capacity is a function of the total surface area available to the PNP molecules and is also a function of the type of activation treatment to which the carbon has been subjected. Many activated carbons have comparable surface areas and pore size distributions, yet do not exhibit similar PNP adsorption capacities.

The PNP isotherm can usually be represented by the Freundlich adsorption isotherm relation.[15, 26] The Freundlich relationship is given by

$$S = aC^n \qquad (1)$$

where

S = the adsorption capacity,
C = the solute concentration, and
a and n ($n < 1$) = arbitrary constants.

An isotherm that can be described by this empirical relationship yields a straight line when plotted on log-log coordinates. The isotherms of Figure 1 have been replotted on log-log coordinates in Figure 2. The slope of this straight line is n; this slope can also be used in characterizing the adsorption behavior of the carbon. A high adsorption capacity at a given PNP concentration, combined with a low value for n, indicates that the carbon will retain more of that capacity at a lower PNP concentration than it would if it had a larger value of n.

Rate of Adsorption

Digiano and Weber [20] have examined the adsorption kinetics of granular carbons and the mechanisms controlling these kinetics. Adsorption of organics by granular activated carbon consists of four steps:

1. Mass transport (diffusion) from the agitated bulk solution across the stationary "film" around the carbon particle—film diffusion;
2. Diffusion through the pores of the carbon to an adsorption site—pore diffusion;
3. The adsorption reaction itself, on reaching the adsorption site—chemical kinetics; and
4. Diffusion of desorbed solute away from the adsorption site. Film diffusion is determined by particle geometry and is not a variable that requires evaluation. The chemical kinetics of the adsorption reaction are much faster than any of the diffusion processes, and the desorption rate is mass-transport limited, as is the adsorption rate. The parameter to be evaluated is the second one, pore diffusion. Naturally, a particle with extremely fine (10 to 20 Å) pores will exhibit a lower rate of adsorption than one with a network of increasingly larger pores, from 10 to 20 Å up to

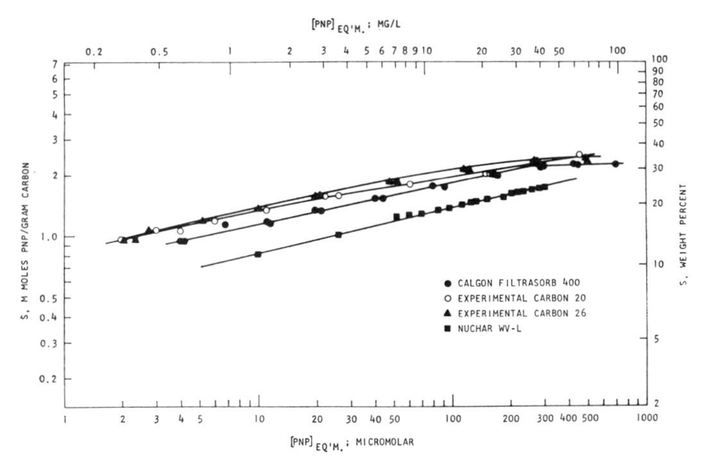

FIGURE 2.—Log-log representation of data from Figure 1. Data for experimental carbon 20 and commercial carbon B plot as straight lines, conforming to Freundlich isotherm, but, as shown by data for the other carbons, such behavior cannot be assumed a priori.

TABLE II.—Rate Constants for *p*-Nitrophenol Adsorption

Carbon	$t_{\theta=0.5}^{1/2}$ (min$^{1/2}$)	PNP $_{\theta=0.5}$ (μM)	k (1/mole-min$^{1/2}$)
Experimental	6.20	132	1,220
Commerical A*	10.2 (11.9)	166 (151)	590 (557)
Commercial C	14.9	130	517
Commerical D	15.8	137	462
Commercial E*	13.1 (21.2)	202 (124)	377 (380)

* Two experiments at different starting concentrations of PNP; the data in parentheses go with the data shown in Figure 3.

several microns, which will provide for rapid access to the interior of the particle.

Figure 3 illustrates the rate of PNP adsorption for one experimental and four commercial activated carbons. These data are presented as the fraction of equilibrium surface coverage attained as a function of the square root of time. The procedure used for obtaining these data is outlined below.

1. PNP (3.5 l, pH 2, 0.3 m*M*) is placed in a 4-l glass vessel, which in turn is contained in a 25.0°C constant-temperature bath. This is allowed to stand overnight to allow equilibrium to be obtained with the glass walls.

2. The test carbon is sieved to 16 × 20 (U.S. Sieve) mesh, yielding a particle size range from 0.841 to 1.19 mm. The solution is stirred at a constant speed (that is 1,200 rpm using a stirrer that will agitate the entire solution volume), and 100 mg of the test carbon is added. Five-milliliter samples can be withdrawn with a pipette, taking care not to include any carbon particles. Samples are taken after about 5, 15, 30, 60, 90, 120, 180, 240, 300, and 400 min, with a final sample taken after 24 hr. The 24-hr point serves as the "equilibrium capacity" value. The samples are analyzed for PNP concentration, and the fractional PNP uptake is plotted versus $t^{1/2}$.

For the curves shown in Figure 3, it is significant to look at the amount of time required to reach an uptake corresponding to one-half the "equilibrium," or 24-hr, surface coverage and to express the adsorption kinetics simply as:

$$\theta = k[PNP]t^{1/2} \qquad (2)$$

where k is obtained for $\theta = 0.5$. Table II summarizes the data for the curves shown in Figure 3. The rate constants differ by a factor of 3.2 from the experimental carbon illustrated to commercial carbon E. The difference in the capital investment required for a physical-chemical treatment plant would be significant for such a difference in adsorption rates. Cover and Wood[5] show that, for a 10-mgd (37,850-cu m/day) tertiary treatment system, the lower capital costs required with a kinetically superior carbon would be reflected in the total operating costs (including amortization) to the extent of a 15 percent reduction in these operating costs per change of a factor of two in contact time (rate). Cover and Wood[5] also show that a 30 percent decrease in contact time (increase in flow rate) results in a 14.5 percent decrease in fixed capital investment for a 10-mgd (37,850-cu m/day) tertiary facility. The adsorption rate parameter is obviously far more important than the activated carbon companies admit, as shown by a lack of any mention of this property in their literature.

It is obvious that the effect of film diffusion is constant for a given particle size and also that film diffusion is not going to be rate-limiting in a high-flux activated carbon adsorber. For a low-flux, downflow carbon adsorber, film diffusion and "active transport" by biological slimes may well be rate limiting. If biological slimes are not allowed to build on the carbon, pore diffusion is still likely to be the rate-limiting step.

Resistance to Attrition

The ability of a granular carbon particle to resist mechanical and hydraulic attrition plays the most important economic role in selecting the

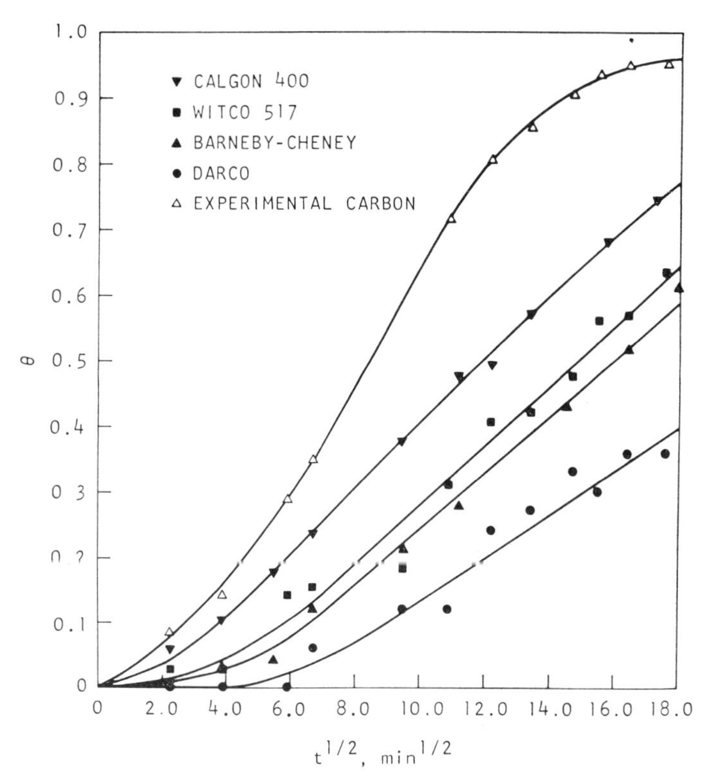

FIGURE 3.—Fractional uptake of *p*-nitrophenol as a function of square root of time.

carbon that will result in the lowest operating costs. Carbon must be regenerable for these systems to work, and the regeneration losses must be kept to a minimum. Juhola and Tepper[12] found that the minimum loss to be expected by burnoff in the regeneration process itself is about 3 percent. Regeneration losses experienced in field tests are found to be from 5 to 20 percent, including mechanical attrition, for 8 × 30 and 12 × 40 mesh carbons. There is obviously some room for improvement in regeneration techniques, but that is not the point of this discussion. Hydraulic attrition losses can be measured in laminar upflow columns; however, experience in this laboratory indicates that mechanical attrition is a far more serious problem than hydraulic attrition. Mechanical attrition is usually measured by a process of shaking a given amount of carbon on a sieve in the presence of 20 to 50 steel balls of one or two sizes. Results are difficult to reproduce, and

TABLE III.—Mechanical Attrition Test Results

Carbon Sample No.	Percent Abraded Average of 5
1	23.1 ± 4.8
2	10.3 ± 2.3
3	14.5 ± 3.3
4	32.9 ± 3.7
5	20.4 ± 2.3*
6	33.6 ± 4.7†

* Average of 4.
† Average of 3.

the extent to which conditions found in screw conveyers and other such devices can be simulated is debatable. Some typical results, however, are shown in Table III. These results exhibited a large amount of scatter, but can be used in a gross fashion to pick out a harder product rather than one that breaks up under the least wear.

For a test such as this, it is reasonably easy to establish a fixed testing procedure in each laboratory and to test the carbons individually.

Summary

The technical evaluation of activated carbons would benefit from the addition of procedures for the determination of adsorption rate and capacity, as well as a determination of the ability of each carbon to withstand a specified level of mechanical attrition. By judicious selection of a material based on these criteria, as well as the price, of course, the operating costs of physical-chemical and tertiary treatment may be reduced by a significant amount.

References

1. Joyce, R. S., and Hager, D. G., U. S. Patent 3,455,820, Calgon Corp., July 15, 1969.
2. Bishop, D. F., *et al.* "Study of Activated Carbon Treatment." *Jour. Water Poll. Control Fed.*, **39**, 188 (1967).
3. Weber, W. J., Jr., "Fluid-Carbon Columns for Sorption of Persistent Organic Pollutants." In "Advances in Water Pollution Research." *Proc. 3rd Intl. Conf. Water Poll. Res.*, Water Poll. Control Fed., Washington, D. C., **Vol. 1**, 253 (1967).
4. Stephan, D. G., and Schaffer, R. B., "Wastewater Treatment and Renovation Status of Process Development." *Jour. Water Poll. Control Fed.*, **42**, 399 (1970).
5. Cover, A. E., and Wood, C. D., "Optimizing an Activated Carbon Wastewater Treatment Plant." *Ind. Water Eng.*, **7**, 4, 21 (1970).
6. Weber, W. J., Jr., *et al.*, "Physicochemical Treatment of Wastewater." *Jour. Water Poll. Control Fed.*, **42**, 83 (1970).
7. Zuckerman, M. M., and Molof, A. H., "High Quality Reuse Water by Chemical-Physical Wastewater Treatment." *Jour. Water Poll. Control Fed.*, **42**, 437 (1970).
8. Weber, W. J., Jr., "Discussion" of "High Quality Reuse Water by Chemical-Physical Wastewater Treatment." *Jour. Water Poll. Control Fed.*, **42**, 456 (1970).
9. "Activated Carbon for Wastewater Treatment?" Publ. D-110, Atlas Chemical Industries, Inc., Wilmington, Del. (1969).
10. "Evaluation of Granular Carbon for Wastewater Treatment." Publ. D-111, Atlas Chemical Industries, Inc., Wilmington, Del. (1969).
11. Product Bull. 20–2a, Application Bull. 20–4, 20–5, 20–9, Calgon Corp., Water Management Div., Pittsburgh, Pa.
12. Juhola, A. J., and Tepper, F., "Laboratory Investigation of the Regeneration of Spent Granular Activated Carbon." Rept. No. TWRC-7, FWPCA, Washington, D. C. (1969).
13. Cover, A. E., and Pieroni, L. J., "Appraisal of Granular Carbon Contacting, Phases I and II." Rept. No. TWRC-11, FWPCA, Washington, D. C. (1969).
14. Cover, A. E., and Wood, C. D., "Appraisal of Granular Carbon Contracting, Phase III." Rept. No. TWRC-12, FWPCA, Washington, D. C. (1969).
15. Mattson, J. S., and Mark, H. B., Jr., "Activated Carbon: Surface Chemistry and Adsorption from Solution." Marcel Dekker, New York, N. Y. (1971).
16. Hunter, J. V., and Heukelekian, H., "The Composition of Domestic Sewage Fractions." *Jour. Water Poll. Control Fed.*, **37**, 1142 (1965).
17. Painter, H. A., and Viney, M., "Composition of a Domestic Sewage." *Jour. Biochem. Microbiol. Tech. Eng.*, **1**, 143 (1959).
18. Snoeyink, V. L., and Weber, W. J., Jr., Tech. Publ. No. T-68-1, Dept. Civil Eng., Univ. of Michigan, Ann Arbor (1968).
19. Keinath, T. M., and Weber, W. J., Jr., Tech. Publ. No. T-68-2, Dept. Civil Eng., Univ. of Michigan, Ann Arbor (1968).
20. Digiano, F. A., and Weber, W. J., Jr., Tech. Publ. No. T-69-1, Dept. Civil

Eng., Univ. of Michigan, Ann Arbor (1969).
21. Mattson, J. S., Ph.D. thesis, Water Resources Science, Univ. of Michigan, Ann Arbor (1970).
22. Mattson, J. S., *et al.*, "Surface Chemistry of Active Carbon: Specific Adsorption of Phenols." *Jour. Colloid & Interface Sci.*, **31**, 116 (1969).
23. Snoeyink, V. L., *et al.*, "Sorption of Phenol and Nitrophenol by Active Carbon." *Environ. Sci. & Technol.*, **3**, 918 (1969).
24. Mattson, J. S., *et al.*, "Surface Oxides of Activated Carbon: Internal Reflectance Spectroscopic Examination of Activated Sugar Carbons." *Jour. Colloid & Interface Sci.*, **33**, 284 (1970).
25. "Basic Concepts of Adsorption on Activated Carbon." Calgon Corp., Pittsburgh, Pa.
26. Weber, W. J., Jr., and Morris, J. C., "Equilibria and Capacities for Adsorption on Carbon." *Jour. San. Eng. Div., Proc. Amer. Soc. Civil Engr.* **90**, SA3, 79 (1964).

ROTATING DISKS WITH BIOLOGICAL GROWTHS PREPARE WASTEWATER FOR DISPOSAL OR REUSE

Wilbur N. Torpey, H. Heukelekian, A. Joel Kaplovsky, and R. Epstein

The purpose of this paper is to report on the development of a system of treatment of wastewater designed to produce different quality effluents suitable for disposal into receiving bodies of water or for reuse. The system has three component parts that can be used either separately or sequentially. The wastewater is treated first on a series of rotating disks with attached biological growths. Depending on the number of sequential units used, it is possible to produce an effluent with successively greater degrees of removal of carbonaceous organic matter which may or may not be followed by the oxidation of ammonia to nitrates. The effluent from this component of the system is then treated on a series of rotating illuminated disks on which is generated attached algae for the purpose of the removal of nutrients. Finally, if so desired, the wastewater having been so prepared can be treated by activated carbon for the removal of residual biodegradable and nonbiodegradable organic materials. The effluent from this treatment then can be reused for certain purposes or made potable through demineralization and disinfection. The system of treatment can be tailored accordingly to a wide spectrum of needs demanded by local situations.

This conceptual system has been developed on the basis of pilot-plant work located at the Jamaica water pollution control plant in New York City.

Removal of Carbonaceous Matter and the Oxidation of Ammonia

The first component is comprised of 10 sequential stages (Figures 1 and 2). Each stage consists of a horizontal shaft on which are mounted forty-eight 3-ft diam (0.9-m) aluminum disks, 0.06 in. (0.15 cm) thick, and spaced on 0.5-in. (1.27-cm) centers. Each shaft is driven separately by a hydraulic motor. The rotation alternately submerges the attached biological growths and exposes them to air. The flow has been 7.5 to 9 gpm (0.47 to 0.57 l/sec) of primary effluent, which results in a theoretical detention time of 5 to 6 min in each stage. The shafts are rotated at an average speed of 10 rpm and are generally operated in a direction opposing the flow of wastewater.

Samples of influent to the biological unit and effluents from each stage were taken from 3:00 to 6:00 PM during the higher load conditions. Initially, the samples were settled and the supernatant was composited for analysis. Subsequently the samples were filtered through a 35-μ microstrainer.

The following determinations were made regularly: biochemical oxygen demand (BOD), chemical oxygen de-

FIGURE 1.—Biological rotating disks, general arrangement.

mand (COD), suspended solids (SS), NH_3-N, NO_2-N, and NO_3-N. Additional tests were made for special purposes. Allylthiourea was used to suppress nitrification in the BOD tests in stages manifesting nitrification.

The paper presents only the results of the initial period of operation from July to November 1969. The temperature of the wastewater during this period ranged between 62° and 78°F (16.6° and 25.6°C).

FIGURE 2.—Close-up of biological rotating disks.

The average results of 33 samples of the operation of the biological units are presented in Table I. The average BOD of the influent (the effluent of the primary tanks from the Jamaica plant) was 124 mg/l which was lowered by Stage 1 to 82 mg/l, thus effecting a reduction of 34 percent in approximately 5 to 6 min of contact time. In Stage 2 it was further lowered to 59 mg/l for an additional 28 percent reduction. An effluent with 19 mg/l of BOD was obtained at the end of Stage 5 for an overall reduction of 85 percent in 25 min of contact time. The effluent from Stages 9 and 10 had a BOD of only 9 mg/l for an overall reduction of 93 percent, exclusive of primary treatment.

The COD was lowered from 303 mg/l to 220 mg/l in Stage 1 for 27 percent reduction. At Stage 5 it decreased to 103 mg/l for 66 percent removal, while the overall removal after Stage 10 was 78 percent. The reason for the lower rates of COD reduction in comparison with the BOD must be ascribed to the presence of

TABLE I.—Performance of the Biological Units*

Parameter	Number of Samples	Influent	Stage Number									
			1	2	3	4	5	6	7	8	9	10
BOD (mg/l)	33	124	82	59	44	28	19	17	14	12	9	9
COD (mg/l)	33	303	220	174	152	121	103	94	85	81	73	64
SS (mg/l)	33	107	69	47	44	31	20	16	14	13	12	9
NH_3-N (mg/l)	31	13	14.2	14.0	14.4	13.6	13.2	12.8	11.0	8.9	6.9	5.7
$NO_2 + NO_3$-N (mg/l)	27	—	—	—	—	—	1.0	2.2	4.3	6.7	8.6	10.4

* Average results for period from July to November.

nonbiodegradable materials. The BOD and COD concentrations in the effluents from each unit are presented in Figure 3.

The average SS in the influent to Stage 1 was 107 mg/l, which was reduced to 69 mg/l in the effluent from that stage for a 37 percent reduction (Table I). After Stage 5 the SS were reduced to 20 mg/l for a reduction of 87 percent. The effluent from Stage 10 had 9 mg/l of SS, giving an overall reduction of 91 percent, exclusive of primary treatment. The SS settled readily and filtered through the microstrainer with increasing rapidity after each successive stage.

The NH_3-N at the first stage increased slightly over the influent because of the hydrolysis of organic nitrogen in the biological growths. At Stages 4 and 5 a slight decrease was obtained. Thereafter, the decrease was more rapid down to 5.7 mg/l at Stage 10. Nitrification started at Stage 5 and increased thereafter by about 2 mg/l at each stage for a total of 10.4 mg/l of combined oxidized nitrogen

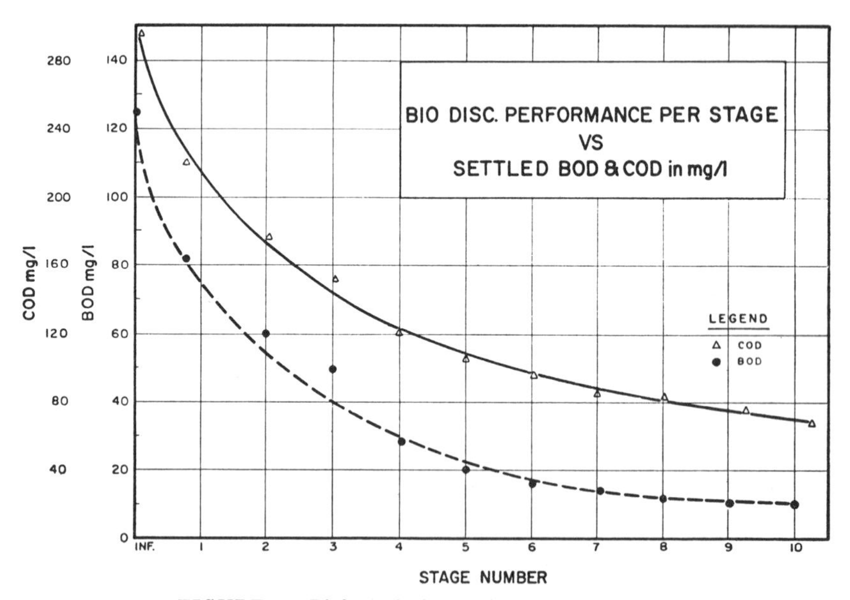

FIGURE 3.—Biological disk performance per stage.

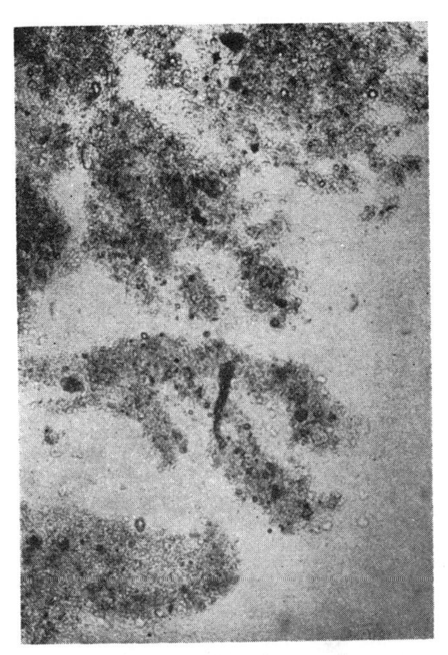

FIGURE 4.—*Zoogloea ramigera.*

FIGURE 6.—Zoogloea, carchesium, and *Sphaerotilus.*

within 30 min of contact time. Higher concentrations were obtained during the warmer months.

It is apparent from these results that rapid nitrification takes place with specialized and established flora under optimum environmental conditions.

Microscopic examinations of the biological growths revealed a succession of different types of microorganisms

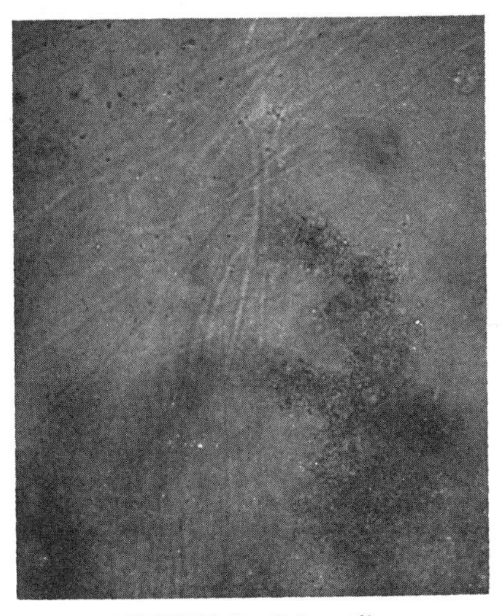

FIGURE 5.—*Sphaerotilus.*

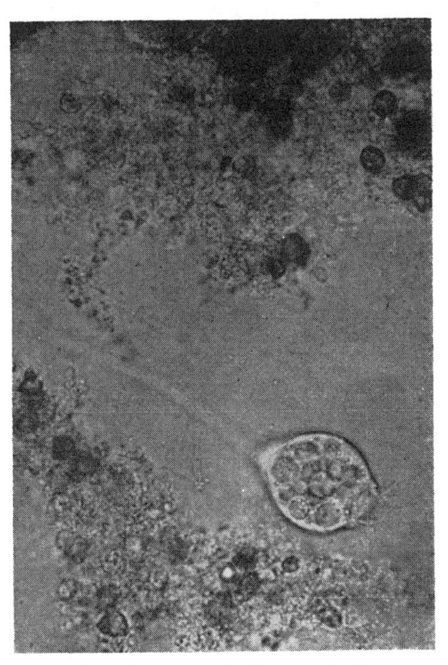

FIGURE 7.—*Vorticella* and *Difflugea.*

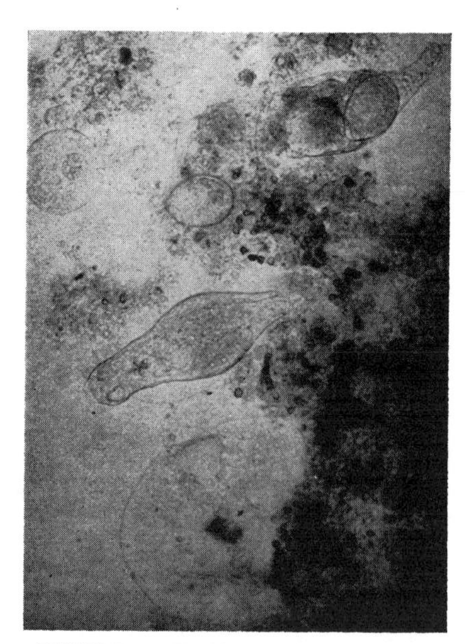

FIGURE 8.—Rotifers, *Difflugea*, amoeba, and ciliate.

starting with a predominance of zoogloeal bacteria and *Sphaerotilus* in the first three stages, followed by an abundant and diversified fauna consisting of free-swimming and stalked protozoa, rotifers, and nematodes in the subsequent stages. The activities of the abundant and diversified microfauna played a major role in the high degree of clarification of wastewater and the destruction of substantial amounts of organic matter (Figures 4, 5, 6, 7, and 8). The extent of the effects of animal predation became evident in the last four stages which, at times, resulted in bare spots on the disk surfaces, the magnitude of which depended on the relative rates of growth of predators versus bacterial slimes. It was apparent that, in contrast with the activated sludge process, a succession of morphologically and biochemically specialized microorganisms developed in the various stages in step with the changes in the substrate composition which resulted in a high efficiency of treatment.

The accumulation of biological growth in the first three stages was more rapid than on each of the succeeding stages. It was found advantageous to remove the growths to prevent bridging and anaerobic conditions in the attached part of the growth. The growths from Stages 1 and 2 were removed once in 4 or 5 days and with decreasing frequency thereafter down to Stage 6. The succeeding stages were cleaned only once in several months. The cleaning in the pilot plant was accomplished by using a water jet. Thus, none of the biological growths was allowed to reach a thickness greater than 1.5 mm.

After cleaning, a biological growth was restored on the disk surfaces of Stage 1 after about 18 hr with the reestablishment of normal efficiency of

FIGURE 9.—Algal unit, general arrangement.

FIGURE 10.—Rotating disk with algae growth.

treatment. The time required for the establishment of growths increased only slightly down to Stage 3. Beyond this stage increasing time was required for the establishment of growth.

Dissolved oxygen (DO) was absent or was present in less than 1 mg/l concentration in Stages 1 and 2. Thereafter it increased progressively to a range of about 4 to 6 mg/l in Stage 10.

The pH values increased slightly to a range of 7.3 to 7.6 up to Stage 6 and then decreased to a range of 7.1 to 7.5 at Stage 10. At the same time, the bicarbonate alkalinity decreased as a result of the production of nitric acid.

The chlorine demand decreased progressively from an initial value of 17 mg/l to 3 mg/l at Stage 6 and to 1 mg/l at Stage 10.

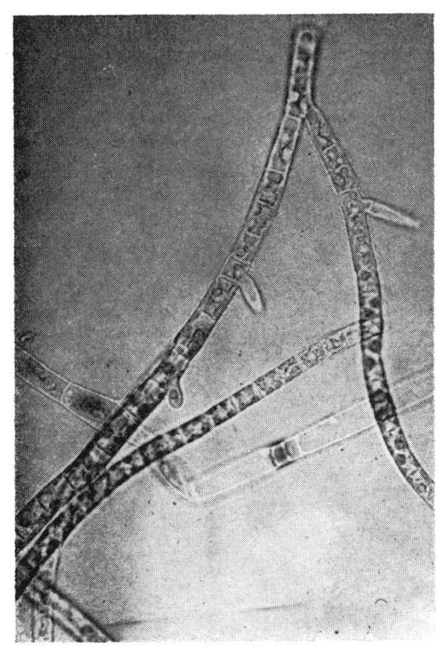

FIGURE 11.—Filamentous algae—hormidium.

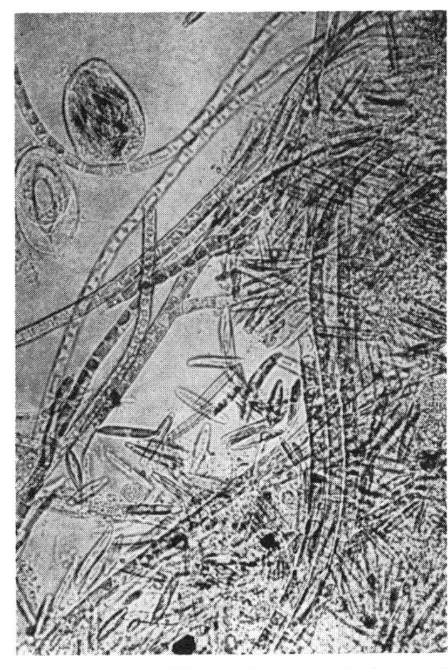

FIGURE 12.—Nematode egg, *nitschia*, and hormidium.

Surfactants were not attacked by the growths in Stages 1 and 2, but thereafter they decreased from an initial value of about 9 mg/l to 1 mg/l at Stage 8, at which stage no further decrease occurred.

Removal of Nutrients

The algal unit (Figure 9) consisted of six stages of partially submerged rotating disks with a triangular cross section. The number of disks on the shafts decreased from 12 in Stage 1 to 2 in the Stage 6. The disks were constructed of 0.06-in. (0.15-cm) thick aluminum, 3 ft (0.9 m) in diam, and were hollow. The disks were exposed alternately to an overhead light source of fluorescent lights enclosed in a hood and immersed in the flow.

Filamentous algae grew only along the outer rims of the disks because only in that area was the light intensity adequate. Exposure of an outer disk to 1,000 ft-c (10,764 lumen/sq m) of illumination from cool white fluorescent tubes produced a luxuriant growth of filamentous algae (Figures 10, 11, 12, and 13). On the basis of a number of observations with varying light intensities, it is planned to replace the triangular disks with flat disks parallel to each other and with sufficient space between them for the insertion of a light source of proper intensity and quality.

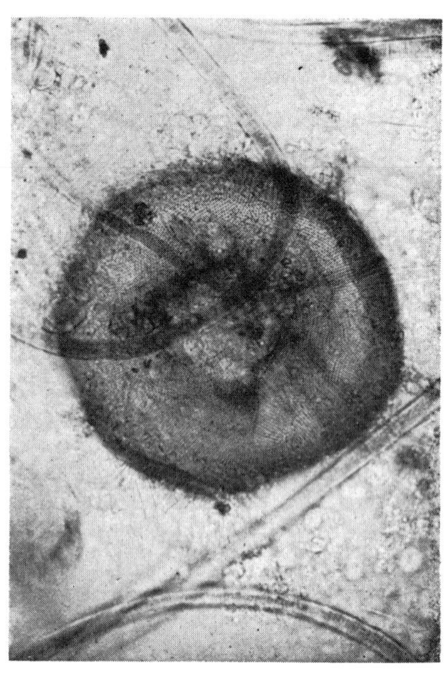

FIGURE 13.—*Oscillatoria* and *arcella*.

Adsorption on Activated Carbon

Carbon adsorption treatment of the effluent from the algal unit was practiced during the last 2 months. About 27 lb (12.3 kg) of virgin granular carbon (12 × 40 mesh) was placed in six packed bed columns 5.83 ft (1.75 m) long and 3 in. (7.6 cm) in diam, each providing for a bed expansion of about 50 percent during the backwashing operation. The hydraulic load-

TABLE II.—Performance of Carbon Columns

Influent	TOC* (mg/l)	Dissolved Organic Carbon* (mg/l)	Color† (Amts.)	Turbidity† (JTU)	Detergent† (mg/l)
	10.0	9.6	17.5	2.0	0.1
Column 1	5.7	4.0	—	—	—
2	3.4	2.6	—	—	—
3	2.6	1.4	—	—	—
4	2.0	1.2	—	—	—
5	1.9	1.0	—	—	—
6	2.1	0.9	2.5	0.9	0.04

* Average of 5 samples.
† Average of 2 samples.

ing rate was maintained at about 5 gpm/sq ft (204 l/min/sq m), and the six columns operated in series.

Preliminary analytical results indicated that the dissolved total organic carbon (TOC) in the effluent from the algal unit varied from 8 to 12 mg/l, while the effluent from the first carbon column varied from 3 to 5 mg/l (Table II). The effluent from Carbon Columns 3, 4, 5, and 6 varied from 1 to 2 mg/l. The results to date did not indicate a significant rise in the TOC leaving Column 1, although some 15,000 gal (57 cu m) had passed through 5 lb (2.3 kg) of carbon in that column, and it was not possible to determine the exhaustion rate.

The pressure across Carbon Column 1 increased at a rate of about 4 lb (1.8 kg) in 24 hr. A daily backwash schedule was practiced to keep the pressure at low levels. In order to reduce the increase in pressure further, a mixed-media filter was installed. As a result the increase of pressure was reduced to 1 lb (0.5 kg) in 24 hr.

Based on these preliminary findings, it seems that the biological treatment, as practiced, was capable of preparing the wastewater for carbon adsorption to such a degree as to make it possible to remove substantially all the organics in the wastewater.

Summary

A method of treatment of primary effluent by a series of rotating disks with attached growths has been developed capable of producing removals of carbonaceous BOD up to 95 percent and the oxidation of ammonia to nitrates. The removal of N and P from the effluent of these units is being attempted by promoting the growth of attached filamentous algae on illuminated rotating disks that are readily harvestable in contrast with the removal of planktonic algae grown in oxidation ponds. The effluent thus prepared is highly amenable to adsorption on activated carbon such that the leakage was held between the limits of 1 to 2 mg/l.

Acknowledgments

This investigation was supported by a grant from the Environmental Protection Agency (WP-17010 EBM). The cooperation and assistance given by Commissioner M. Feldman and Assistant Commissioner M. Lang of the Environmental Protection Administration of New York City is gratefully acknowledged.

TREATMENT OF COLD-MILL WASTEWATER BY ULTRA-HIGH-RATE FILTRATION

C. R. Symons

Bethlehem Steel Corporation is engaged in a corporation-wide program to meet the recently upgraded water quality criteria for effluents from plant operations into receiving waters. The wastewater effluents from cold reduction rolling mills contain concentrations of emulsified oil and suspended solids (SS) higher than those permitted by water quality criteria. At the Lackawanna (N.Y.) plant, five distinctly different methods were proposed for treating this wastewater, and in order to compare the feasibility of these processes, pilot-scale tests were made of each. The results demonstrated that the most practical method for the conditions at Lackawanna is ultra-high-rate (UHR) filtration at a rate of 8 to 12 gpm/sq ft (326 to 488 l/min/sq m) of filter area.

The Lackawanna plant, located south of Buffalo, N. Y., on the shore of Lake Erie, is the second largest integrated plant in the corporation and the fourth largest in the country. Lackawanna produces 2 mil tons (1.8 mil metric tons) of flat-rolled steel products annually. Both the 54- and 75-in. (137.2- and 190.5-cm) cold reduction mills at Lackawanna are 4-stand, 4-high tandem mills. Most of the cold-mill product is sold to automobile manufacturers.

After pickling to remove an oxide film, coils from the hot mills are oiled to provide lubrication during cold reduction and protection during storage prior to rolling. The oil used is about 1.4 lb/ton (0.6 kg/metric ton) of steel rolled and ends up in the cold-mill wastewater. These two cold mills use 3,000 gpm (189 l/sec) of water to clean the sheet and carry away heat generated during cold rolling. A detergent is often added at the last stand to remove all traces of oil from the sheet.

Figure 1 is a flowsheet of the water circuit at the cold mills. Wastewater from the mills enters a collecting sump and contains both floating and emulsified oils in concentrations up to 700 mg/l and SS up to 250 mg/l. The floating oil is removed by skimming and, during the pilot-scale test period, the emulsified oil concentration of the wastewater averaged 230 mg/l and the SS, 100 mg/l. New York State establishes a maximum allowable limit of 15 mg/l oil and 40 mg/l SS in industrial effluents discharged to boundary waters but also specifies that the oil content of industrial effluents should never exceed amounts "as to be injurious to edible fish or shellfish."

The objective of the test work at Lackawanna was to produce an effluent containing less than 15 mg/l of oil.

Treatment Methods Evaluated

The methods considered for treating this oily wastewater were:

1. Acid chemical treatment with cold lime neutralization;

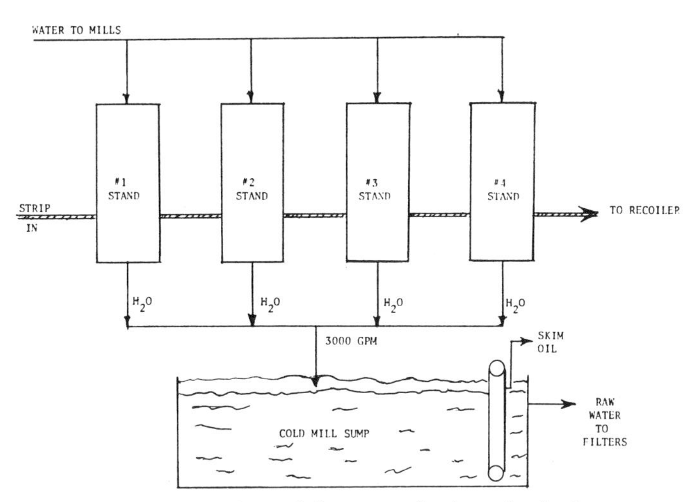

FIGURE 1.—Flow sheet of the water circuit at the Lackawanna cold mills.

2. Solvent extraction in an electrofining precipitator;

3. Alkaline chemical treatment followed by dissolved air flotation;

4. UHR filtration alone or in combination with filtration through diatomaceous earth; and

5. Diatomaceous-earth (DE) filtration in a pressure-precoat filter. All but the acid chemical treatment were pilot tested.

The objective of these tests was to determine which method or combination of methods would produce an acceptable effluent at the lowest capital and operating cost. UHR filtration was shown to be the process that best met these specifications.

Some of the objections to the other processes that contributed to their higher costs were:

1. For the volume of effluent from Lackawanna's cold mills, it was calculated that the acid treatment/cold lime neutralization process would generate daily 80 to 100 tons (72.5 to 90.7 metric tons) of sludge having the consistency of cold cream. This type of sludge is of no value and would cause a disposal problem.

2. A pilot-plant study showed that solvent extraction produces an interface layer that is difficult to treat or dispose of. The effluent from this process would require further treatment, such DE filtration, to conform with the New York State criterion for oil.

3. A pilot study showed that alkaline chemical treatment followed by dissolved air flotation requires a high pH and extremely large quantities of alkali for good floc formation. Control of pH is a problem because of fouling of the pH electrodes. Also, carbonate precipitation occurs after alkali addition and causes plugging of pipelines and pumps.

4. DE filtration on a pilot scale produced an excellent effluent. However, daily body feed costs, which are proportional to the inlet oil concentration, are considered excessive. These costs could be minimized if a pretreatment method were developed to reduce the inlet oil concentration. As a matter of fact, UHR filtration tests made previously at Lackawanna and also on refinery wastes in California indicated that UHR filtration would be effective for this purpose.

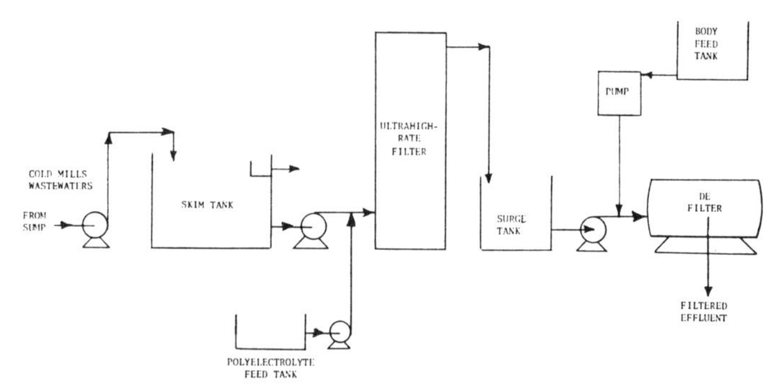

FIGURE 2.—Ultra-high-rate—diatomaceous-earth filtration plant for treating wastewater from Lackawanna cold mills.

UHR Filtration Process

The pilot-scale test program for the development of an effective procedure for a combined UHR-DE filtration process at the two cold reduction mills comprised two phases. In the first phase, consisting of 14 tests, the main features of a basic UHR procedure were established.

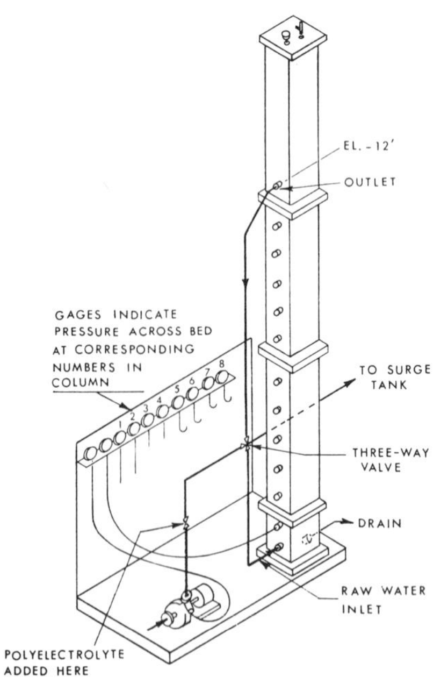

FIGURE 3.—Isometric sketch of ultra-high-rate pilot filter. (In. × 2.54 = cm.)

The features were (*a*) the use of steam during backwashing, (*b*) the proper type and depth of filter beds, and (*c*) an optimum rate of 8 gpm/sq ft (326 l/min/sq m).

The second phase of the program, consisting of 43 tests, included (*a*) testing the overall effectiveness of the UHR procedure to find out in particular whether oils would build up in the filter media, (*b*) developing an effective procedure for treating the backwash water, and (*c*) testing the effectiveness of possible additional steps, such as the use of coagulant, to reduce further the average effluent oil concentration.

Phase 1

Figure 2 is a flowsheet of the pilot plant during the first phase of the program. After passing through a skimmer tank, waters from the common collecting sump for both mills were first filtered through a 1-sq ft (0.09-sq m) UHR filter and then through a pressure-precoat DE filter.

UHR Filter

The UHR pilot filter* is shown in Figure 3. A unique feature of this unit was that it had a transparent front that permitted observation of the action of the filter bed during filtering and backwashing. All the UHR filter

* De Laval Turbine Inc., Florence, N. J.

tests made with this filter pertinent to the design and installation of the full-scale plant at Lackawanna were done by filtering upflow. The filter was backwashed in the same upflow direction.

Filter Media

For the filter media a graded gravel and sand filter bed was used with the coarsest material on the bottom. The initial tests were made to determine the optimum composition of the filter bed and the filter backwash sequence. As seen in Figure 4, the filter bed selected consisted of 62 in. (157.6 cm) of gravel topped with 37 in. (94 cm) of sand. Experience indicated that to be effective the gravel should range from a 1.5-in. (3.8-cm) size at the bottom to 0.125 in. (0.32 cm) toward the top of the gravel bed, a total of eight grades being employed. The sand size averaged 2 mm and ranged from 1.7 to 2.4 mm.

Backwashing

Development of a procedure that insured an effective backwash turned out to be the key to the removal of emulsified oil from the cold-mill wastewater by UHR filtration. It was evident early in the test program that the UHR filter was capable of reducing the inlet oil concentration below the 100 mg/l objective, provided the filter bed was clean at the start of each test run. Poor results were obtained if the filter was not backwashed properly between runs.

The normal backwash procedure for UHR filters was to scrub the bed first with air and water and then follow with a water flush. Several variations and combinations of air-water scrubs and water flushes were tried unsuccessfully; that is, after each backwash the filter appeared dirtier than it was at the start of the run. This was caused by a gradual buildup of oil on the filter media. This problem was solved by the use of 130 lb of steam/sq

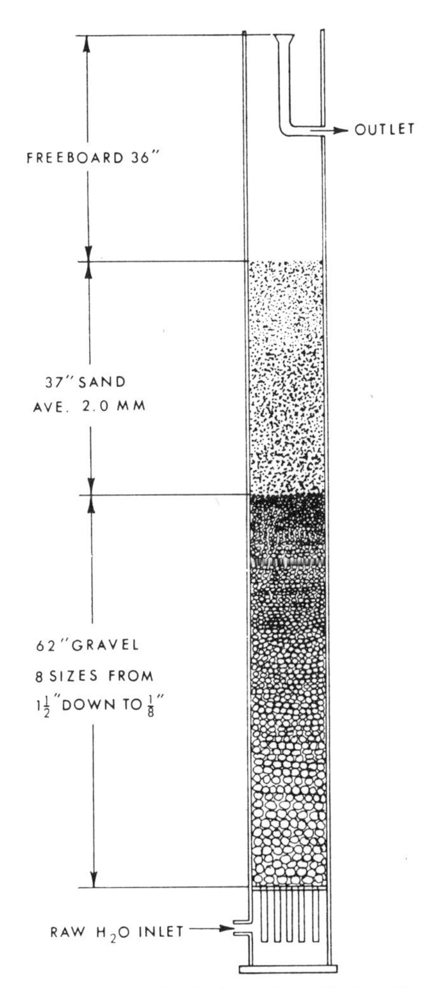

FIGURE 4.—Gradation of media in pilot filter unit. (In. × 2.54 = cm.)

ft (634 kg/sq m) of filter area in the backwashing cycle in addition to the air and water. The efficient release of the oil effected by the steam was attributed to the lowering of the viscosity of the oil adhering to the media. An extensive series of tests was then begun to evaluate UHR filtration using this backwash procedure.

Phase 2

After the basic UHR procedure was established in the first phase of the

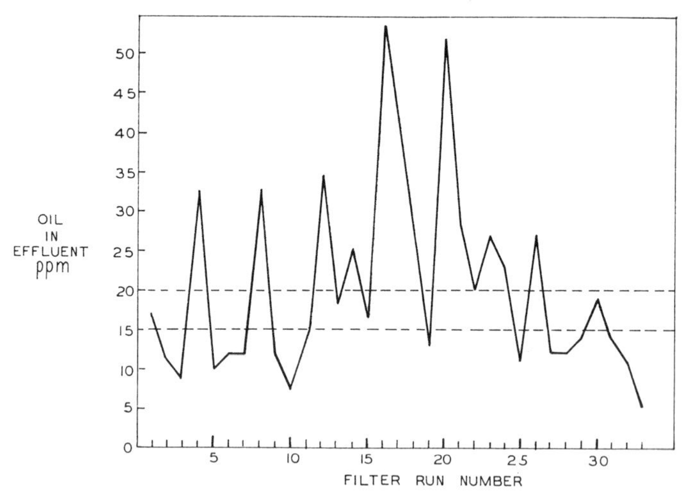

FIGURE 5.—Effluent oil concentration of ultra-high-rate filter tests.

testing program, the filter cycle length was set at 12 hr. Because pressure drop across the filter bed varied between 0.8 and 3.0 psi (0.056 and 0.21 kg/sq cm) and was therefore not a reliable indicator, the filter was backwashed on a timed cycle.

Using the basic UHR procedure, 33 tests were run to check the effectiveness of the process and to find out whether oils would build up in the filter media. Oil buildup would be evidenced by a steady increase in the oil concentration of the filtered water.

The average concentration of oil in the effluent of 33 filter tests using the basic UHR procedure was 20 mg/l. As shown in Figure 5, the concentration exceeded 20 mg/l in 11 runs, was less than 15 mg/l in 16 of the runs, and between 15 and 20 mg/l in the 6 remaining. The average inlet oil concentration was 230 mg/l.

The cold-mill wastewater also contained an average of 100 mg/l SS. This was reduced to an average of 28 mg/l in the UHR filter effluent. The laboratory reported that these solids were principally composed of iron and iron soap, and they had an 80 to 90 percent weight loss when ignited at 1,400°F (759°C).

Treatment of Backwash Water

The first phase of the program had established the importance of backwashing, and one of the purposes of the tests during the second phase was to develop an effective way of treating the backwash water.

In the operation of the UHR filter, backwashing is done in the same upflow direction as filtration. The backwash procedure was designed to drain as much entrapped material as possible through the bottom of the filters.

The volume of backwash water, including all drains and steam condensate, is equal to about 5 percent of the filtered water. This backwash water is returned to a skimmer tank where greater than 90 percent of the oil floats and is skimmed off. After skimming, the backwash water is recycled to the feed inlet of the filter.

The oil skimmed from the backwash tank contains 40 percent water and 5 percent solids and has a gross heating value of 11,000 Btu/lb (6,105 kg-cal/kg). This oil was found to be compatible with plant fuel oil in concentrations up to 10 percent. The heating value will therefore be utilized when disposing of the skimmed oil.

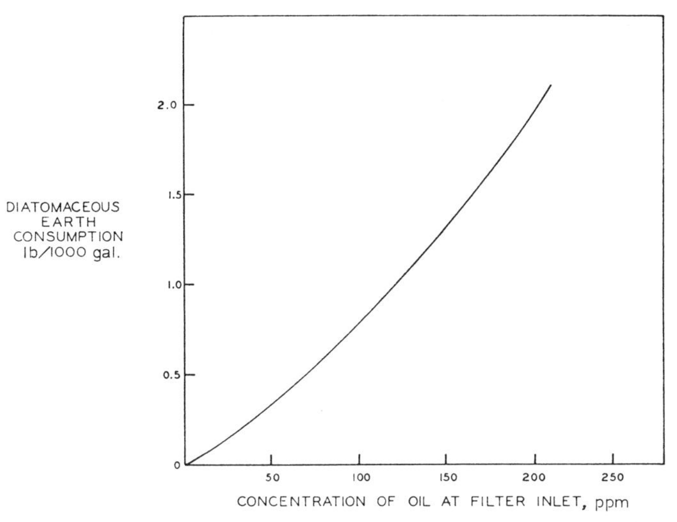

FIGURE 6.—Diatomaceous earth body feed consumption for various filter inlet oil concentrations. (Lb/1,000 gal × 0.00012 = g/cu m.)

Additional Steps

The basic UHR procedure established in the first phase had succeeded in bringing the average effluent oil concentration down to 20 mg/l. Additional treatment steps to improve on the 20-mg/l average result were considered. To possible methods available for this purpose were adding a coagulant to the filter influent or filtering in a DE filter.

Addition of Coagulant

Bench-scale tests had indicated that the addition of a polyelectrolyte coagulant would improve filter effluent quality. Pilot tests made in the UHR filter showed that the addition of 5 mg/l of polyelectrolyte† at the filter inlet reduced the effluent oil concentration to about 10 mg/l and the SS concentration to 17 mg/l. The filter media also seemed easier to backwash when polyelectrolyte coagulants were used.

DE Filter Tests

The DE filter used with the UHR filter was a pressure-precoat leaf filter having an area of 8 sq ft (0.74 sq m). Tests made with this filter prior to the work with the UHR had indicated that the same intermediate grade of diatomaceous earth was suited for both precoat and body feed. Also, as shown in Figure 6, this work showed that DE body feed consumption varied with the inlet oil concentration.

The UHR filter was expected to provide to the DE filter a uniform feed that would analyze about 100 mg/l of oils but would be free of floating oils. The 20-mg/l effluent actually provided by the UHR filter resulted in lower precoat and body feed consumption for the DE filter than was anticipated. At a filtration rate of 1 gpm/sq ft (40.7 l/min/sq m), the DE effluent oil concentration was 5 mg/l. The precoat weight was 0.1 lb/sq ft (488 g/sq m), and the body feed rate of 0.2 lb/1,000 gal (24 g/cu m) of filtered water.

† Nalco D-2017, Nalco Chemical Co., Chicago, Ill.

TABLE I.—Estimated Capital and Operating Costs for System to Treat 3,000 gpm (189 l/sec) Cold-Mill Wastewater

System Description	Capital ($)	Operating ($/day)
Ultra-high-rate filtration	750,000	300
Ultra-high-rate filtration + coagulant	775,000	372
Ultra-high-rate filtration + diatomaceous earth	1,000,000	440

Conclusion

On the basis of the results of this test work, the plant has constructed a UHR filtration plant to treat the 3,000 gpm (189 l/sec) of cold-mill wastewater. Provision was made for polyelectrolyte addition to the filter inlet water if further reduction in effluent oil concentration is found necessary. A summary of the estimated capital and operating costs for this installation is given in Table I.

UHR filtration, with provision for the addition of a coagulant, was selected as the treatment method for cold-mill wastewater because it was found to be more economical in capital and operating costs than other processes such as chemical treatment flotation and also because the by-product sludges are more suitable for disposal.

As demonstrated by pilot-plant tests, the UHR method of filtration produces a satisfactory effluent and costs less to install and operate than any of the processes tested. Engineering and construction of the UHR filter plant is complete, and the plant began operating in October 1970.

Acknowledgments

Thanks are expressed to De Laval Turbine Inc. personnel as well as to members of the Engineering, Fuel, and Cold Strip Mill departments at Lackawanna for their cooperation during the conduct of the test work.

NEW CONCEPTS FOR MANAGING AQUATIC LIFE SYSTEMS

John Cairns, Jr.

The expanding human population and the problems associated with the increase in numbers of people have amplified the awareness that Earth is a rather small planet with finite resources and a life support system that is threatened by air and water pollution, and by solid wastes. This awareness was heightened by the space program which gave the average citizen a visual and emotional confirmation of what we had already grasped intellectually—namely, that we are on a small, beautiful planet surrounded by a thin, vulnerable atmosphere. At about the same time, the concept of "Spaceship Earth" came into use (1). Earlier, but in the same era, the existence of a variety of "ecological backlashes" was publicized generally by the appearance of Rachel Carson's "Silent Spring." These events are responsible in part for a very marked change in public attitude toward environmental problems in general and toward the role of the ecologist-biologist in particular.

The most important new public expectation, and one critical to the future role of biologists, is the development of a capability for prediction. Engineers can predict that with so much concrete, steel, etc., and with a given amount of labor and money, a dam can be built on a river which will enable them to regulate flow below the dam. The flow figures can be predicted with reasonable accuracy and when the dam is built, the performance is generally within the original estimate. In contrast, biologists have developed only relatively recently rather primitive prediction systems for complex natural environments. Most of these predictions have been restricted to ecological catastrophe and the doom that will follow if certain practices are continued (i.e., the extinction of our national symbol, the bald eagle, by DDT). Though many of these predictions still are contested, there is now substantial evidence that many of them are essentially correct. As a result, decision makers increasingly are including biological evaluations and predictions in their plans and a number of bills before Congress provide evidence that this is becoming standard procedure. This is all to the good! The biologist has an important role to play if this planet is to be maintained viable while our industrial society continues to provide the essentials for the "good life." On the other hand, if biologists are to fulfill their role properly, a whole new set of capabilities must be developed—these form the basis of this paper.

We might start by considering some prerequisites for making full use of natural ecosystems without damaging them.

1. Prediction systems, not only to predict catastrophe but also to predict the consequences of various types and intensities of environmental use.
2. Simulation techniques—the use of scale models to approximate various alternative uses of the environment so

that consequences of all possible combinations and intensities of use can be estimated in the planning stages and the optimal beneficial combination of uses can be made.

3. Rapid biological information systems, both "in-stream" and "in-plant," which will permit environmental quality control to be maintained and will provide an early warning of impending environmental problems.

4. Aquaculture techniques that will enable biologists to restore degraded and damaged areas to a condition that will permit fuller beneficial use than is now possible and will enable biologists to make fuller use of an area where non-biological considerations have decreed that the natural environment must be altered drastically.

5. The identification of both the life support functions of our environment and the requirements of the environmental components essential to the maintenance of these functions. All of the above should be as quantitative as possible!

System Components

Prediction Systems

Bio-assay techniques have been used for many years and recently there has been a very marked trend toward greater sophistication in the techniques being developed and used. For example, bio-assays of industrial wastes have progressed from short-term tests using a single species with death as an end point to long-term tests often involving several species and even communities of organisms. When single species are used in these chronic tests, changes in respiration, growth, reproductive success, electrocardiogram or movement patterns, and other functional changes may replace the use of death as a criterion of response. Changes in structure and other characteristics are used when bio-assays are run with entire communities. If the rate of development of bio-assay techniques in the next twenty years merely equals that of the past twenty years, there is little doubt that a family of suitable predictive techniques will be available. It seems quite likely that the rate of development in the next twenty years will greatly exceed that of the past twenty. There is still an excessive time lag between the development of techniques and their use in practical situations. This is probably due to the isolation of large segments of the academic community from the environmental consequences of economic development and to the failure of some industries and some regulatory agencies to make full use of the capabilities of universities in managing environmental problems.

Simulation Techniques

The development of simulation techniques involving the use of scale models becomes increasingly important as our population grows and more intensive use is made of the finite space available to us. In the past when we damaged an environment seriously, we could move on to a new undamaged environment and avoid most of the immediate consequences of poor management. Perhaps the last big movement of this sort in the United States was the exodus from the Dust Bowl. However, since most of the ecosystems of the United States are at or near tolerable stress levels and many already are past these levels, we no longer can go to virgin territory and escape our environmental mistakes. As a consequence, we can afford fewer mistakes without immediate penalty than we could in the past. One of the obvious protective measures we might take to prevent major ecological or environmental problems is to simulate prospective new uses in scale or laboratory models and restrict most of our mistakes to these. This practice is so common in engineering (for example, the U. S. Army Corps of Engineers river models at the Waterways Experiment Station, Vicksburg, Miss.)

and industrial circles that it would hardly need mention were it not for the fact that ecological scale models or environmental simulation systems are not now commonly used. However, ecologists now are becoming quite interested in developing scale models to simulate various environmental systems and the practice should become increasingly common in the future. Of course these suffer the weaknesses of all scale models and are still in primitive stages of development. They need not be extremely expensive and may be used to generate data which could be useful in preventing large scale mistakes. For example, many of the events which have occurred in Lake Erie could probably have been simulated in models.

Biological Information Systems

The development of rapid biological information systems, both in-stream and in-plant, is essential to the maintenance of adequate environmental quality control once an environmental use plan has been put into practice. We need to know the effects on biological organisms of a waste discharge before it enters the receiving stream as well as the biological effects after it enters the stream, and this information should be produced rapidly. Present systems are much too slow in view of the fact that the constituents of a waste stream are likely to vary from hour to hour and from day to day. Potentially disastrous materials should be detected before they enter a receiving stream if at all possible and at the very least, before substantial damage has been done in the receiving stream itself. Several potentially useful methods for rapid in-plant monitoring are being explored (2) (3) (4) (5) and one rapid in-stream method is now operational (6) (7) (8). The in-plant methods just mentioned use changes in heart rate, breathing signal, and movements of the entire fish within a container to detect sublethal concentrations of toxicants in a waste discharge. If successful, these and other "early-warning" in-plant systems could be used to determine the toxicity of a waste before it left the plant so that the appearance of a harmful concentration of a toxicant would activate a control system and shunt the waste immediately to a holding pond or recycle it for additional treatment. This continual information about the toxicity of a waste should enable sanitary engineers to identify periods of operation likely to produce the most toxic wastes as well as identifying those components of the production process which contribute most of the toxicity. Theoretically, this could be accomplished with standard bio-assays currently used, but rarely are enough samples taken over a sufficient period of time to give the range of information that would be available with continually operating bio-assay techniques; the cost also probably would be much higher. A few industries have had portions of their wastes flowing through fish tanks, but usually the criterion used for change was death rather than some change in function such as respiration rate, heart rate, etc. This is not surprising since the primary purpose of these units seems to have been for public relations rather than data gathering. Full development of useful early warning systems with rapid information feedback will probably take a number of years and will require the close cooperation of a variety of disciplines. No doubt the early developmental period will have its share of failures, but it is highly probable that effective systems can be produced and that their use will substantially improve environmental quality control. Since the ultimate test of the effectiveness of a waste treatment process should be in the receiving stream, in-stream early warning systems also should be developed so that a more or less continual flow of information also can be obtained from at least one point above and several points below a waste discharge pipe. The goal for in-stream rapid biological information systems is the same as that for the in-plant, i.e., to find one or more biological responses that will give early warning of some fundamental change (somewhat comparable to the increased body temperature of a human). Development of in-stream early warning systems probably will require considerably more time and funds than the in-plant systems, primarily because the in-plant systems are completely artificial, while the in-stream techniques use complex natural systems with a greater number of uncontrollable variables. Nevertheless, having an in-stream biological monitoring system that will provide a continual flow of information in minutes or hours is a desirable goal and one to which we should aspire despite the difficulties.

Aquaculture Techniques

The development of aquaculture techniques is in such a primitive stage of development that it offers comparatively few possibilities to an environmental manager at this time. Some relatively new and extremely promising experiments are being carried out (9) (10) although the scale of experimentation is minute compared to the need. Many of our aquatic environments now are being changed so drastically and rapidly that it is unlikely that restoration is possible with the limited resources now available, even when restoration is within our capability. However, it may be possible to develop new strains, new hybrid species, etc., and eventually entire new aquatic communities capable of inhabiting at least a few of these areas and thereby increasing their usefulness. A good example of the potential of this approach may be found in areas near heated wastewater discharges of steam-electric power generating plants. In some cases these may produce temperature increases that make the environment unsuitable to the resident species. Rather than wait for a natural "restocking," species adapted to the new conditions might be introduced deliberately and the usefulness of the area increased. There are several serious drawbacks to this procedure which have been discussed elsewhere (11).

Life Support Functions

Last but certainly not least is the identification of the life support functions of our environment and the determination of the operational prerequisites for those components essential to the proper performance of these functions. For example, the human race is quite dependent on the oxygen in earth's atmosphere, so much so that deprivation for a few minutes is enough to cause death. Recently air pollution problems in some of the large metropolitan areas generated some interest in atmospheric contaminants but as yet there seems to be little concern about the replenishment of atmospheric oxygen. Most people know that carbon dioxide is utilized by plants which produce oxygen but most seem to assume that since this process has been going on for countless years it will continue to do so. However, there is some recent evidence (12) that a persistent pesticide in very low concentrations may interfere with the photosynthetic capability of marine diatoms and an estimated 70 percent of our oxygen supply comes from the oceans. NASA planners would be discharged if they forgot to provide an adequate oxygen supply for the astronauts, yet the inhabitants of Spaceship Earth blithely go on introducing more and more contaminants into the ocean and carrying out other activities which may endanger the oxygen-producing capacity of the earth's life support system. There is as yet only fragmentary evidence that this can happen but the possibility should be enough!

Once the life support functions essential to our survival have been identified and quantified, the ecosystems responsible for these functions can be identified and their operational prerequisites determined. We then can estimate the amount of stress that these systems can be subjected to without interfering with their function, using some of the approaches previously described, and devise means of insuring that appropriate conditions for proper ecosystem function are maintained.

Our problems are different only in order of magnitude from the crew of a space ship on a very long voyage. We want to be certain that our life support system will function properly for the entire length of the cruise. Once survival is assured, we want to know something about the quality of our existence including recreational, aesthetic, nutritional, social, etc., aspects. Our response to environmental problems has hitherto been almost entirely defensive. That is, we have responded primarily to threats and have ignored constructive management and planning. The present condition of our environment gives clearcut evidence that this limited response pattern is not effective and that management and planning are badly needed. We should be able to choose from a a series of options and alternatives the combination best suited to our survival in a diverse and attractive world and to control activities which threaten the implementation of our chosen program. Since we are a society almost compulsively dedicated to change, we are desperately in need of adequate prediction systems that will enable us to determine with reasonable accuracy what the consequences of contemplated changes will be. In order to do this economically and speedily, and to prevent large-scale environmental disasters, we need to be able to simulate in scale models the various alternative uses which might be made of the environment and to estimate what the consequences of these will be. Only then will we be able to have adequate environmental planning. Environmental planning alone, however, will not be effective unless good quality control techniques are developed, as well as adequate environmental management practices. These will require rapid biological information systems so that we get a continuous flow of information on critical biological, chemical, and physical environmental characteristics, enabling us to predict unfavorable changes. Only then will the number of environmental catastrophes, emergencies, and threats be reduced from the present level.

Conclusions

There are a number of questions which must be answered if the proposals in this paper are to be effectively implemented. Does the burden of proof, as to what effects a particular discharge is having on the environment, ultimately rest with the contributor? If so, when will this be enforced and if not, who will have this responsibility? Who should promote and financially support the use of models on systems as complex as Lake Erie? Is it healthy to have one agency or organization generate all the information on complex environmental problems even though it may be cheaper? How can the techniques for "in-plant" monitoring be used for industrial discharges into municipal systems? In large plants where many compounds are released, how might the "in-plant" warning system be modified to indicate which production area is responsible for abnormal conditions.

This paper indicates that the methods and thought patterns of the past will not be adequate to meet the challenge of protecting our ecological life support system without endangering our industrial base.

Acknowledgments

Kenneth L. Dickson, Richard E. Sparks, William T. Waller and Dr. J. Steven Anderson are thanked for comments on a rough draft of this manuscript. These thoughts were developed during a research project sponsored by the Manufacturing Chemists Association.

References

1. Boulding, K. E., "The Economics of the Coming Spaceship Earth." In "Environmental Quality in a Growing Economy." Johns Hopkins Press, Baltimore, Maryland (1966).
2. Shirer, H. W., Cairns, J., Jr., and Waller, W. T., "A Simple Apparatus for Measuring Activity Patterns of Fishes." Water Resources Bull., **4**, 3, 27 (1968).
3. Sparks, R. E., Heath, A. G., and Cairns, J., Jr., "Changes in Bluegill EKG and Respiratory Signal Caused by Exposure to Concentrations of Zinc." *Assoc. Southeastern Biol. Bull.* **16**, 2, 69 (1969).
4. Scheier, A., and Cairns, J., Jr., "An Apparatus for Estimating the Effect of Toxicants on the Flicker Frequency Response of Bluegill Sunfish." *Proc. 23rd Ind. Waste Conf.*, Purdue Univ., Ext. Ser. **132**, Part 2, 849 (1969).
5. Waller, W. T., and Cairns, J., Jr., "Changes in Movement Patterns of Fish Exposed to Sublethal Concentrations of Zinc." *Assoc. Southeastern Biol. Bull.* **16**, 2, 70 (1969).
6. Cairns, J., Jr., Albaugh, D. W., Busey, F., and Chanay, M. D., "The Sequential Comparison Index—A Simplified Method for Non-biologists to Estimate Differences in Biological Diversity for Stream Pollution Studies." *Jour. Water Poll. Control Fed.*, **40**, 1607 (1968).
7. Cairns, J., Jr., Dickson, K. L,. Sparks, R. E., and Waller, W. T., "Biological Monitoring—Progress Report on Manufacturing Chemists Association Research Project." Paper presented at 43rd Ohio Water Poll. Control Conf., Toledo, June 18–20, 1969.
8. Dickson, K. L., and Cairns, J., Jr., "The Use of Basket-type Artificial Substrates for Stream Pollution Monitoring." (Manuscript in preparation).
9. Cook, C. D. K., "Phenotypic Plasticity with Particular Reference to Three Amphibious Plant Species." In "Modern Methods in Plant Taxonomy." Academic Press Inc., London, England, 97 (1968).
10. North, C. W., "Remarks." Natl. Symposium on Thermal Poll.—The Biol. Considerations, Vanderbilt Univ. Press (In Press).
11. Cairns, J., Jr., "Ecological Management Problems Caused by Heated Waste Water Discharge into the Aquatic Environment." Governor's Conf. on Thermal Pollution, Mich. Water Resources Comm., **21** (1969).
12. Wurster, C. F., Jr., "DDT Reduces Photosynthesis by Marine Phytoplankton." *Science* **159**, 1474 (1968).

PRE- and POST-CONSTRUCTION SURVEYS

JOHN CAIRNS, JR., KENNETH L. DICKSON, and ALBERT HENDRICKS

Streams, rivers, or lakes are complicated ecological systems. It is becoming increasingly obvious to the aquatic ecologist and the sanitary engineer that only after extensively studying a receiving system can they give information describing the environmental impact of a waste discharge. Surveys describing the possible effects of a waste discharge upon the receiving system are highly sophisticated. No longer is it possible to approach a stream, collect a few bottom-dwelling organisms, take a water sample for dissolved oxygen and BOD_5, then call this a stream survey. Studies of such aspects as community structure, nutrient recycling, photosynthesis and respiration, sub-lethal toxicity effects, and others show that stream surveys must be performed by a skillfully trained team of competent aquatic biologists and sanitary engineers.

Of utmost importance in a stream survey, however, is acquisition of a strong information base. Only after gathering extensive data on community structure, nutrient recycling, etc., can one diagnose the effects of a waste discharge upon a stream. And one accumulates this strong information base via pre- and post-construction surveys.

Pre-construction Survey

One common problem of water resources development is selection of a project site allowing maximum use without environmental degradation. To get this type of information requires development of a series of prediction systems which allow an ecologist to rank the potential construction sites. Perhaps one of the best ways to obtain ecological information to be used in water resources management is through a pre-construction survey.

The survey should be carried out by a team of chemists, ecologists, engineers, and taxonomists to get a complete picture of the chemical, physical, and biological condition above and below the potential site location. If adequate background data are to be generated, the team should consist of one or more chemists, a bacteriologist, an algologist, a protozoologist, one or more invertebrate zoologists (including an aquatic entomologist), an ichthyologist, and a sanitary engineer. Obviously, such a survey can be moderately expensive. Actual cost would depend on various factors including size and structure of the river, and number of species likely to be encountered. Certainly, the lower Mississippi is a more difficult river to survey than a small stream. And a stream already degraded by pollution is apt to have fewer species present than an unpolluted stream, and will thus be less costly to survey. Before such a survey is contemplated, however, it is well to have a preliminary survey made by a generalist who can make a firm estimate of the costs involved and place reliable estimates as to completion time of the project.

We all know of industries compelled to limit production and expansion in order to install costly waste treatment facilities to help make their discharge compatible with the receiving environment. Pre-construction surveys can provide data enabling an industry to anticipate and interpret its operations to minimize environmental impact. Stream surveys prior to starting operations are important for a number of reasons:

- Only through such surveys can one determine the system's biological, chemical, and physical characteristics

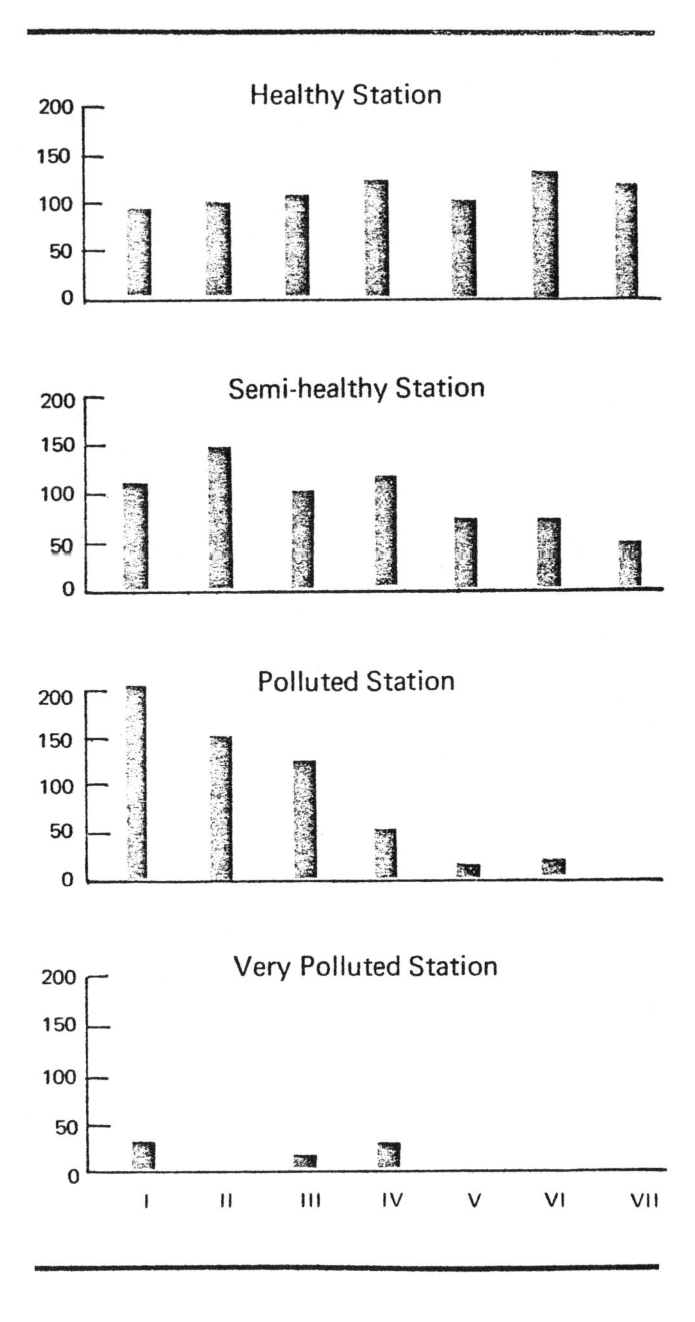

Population structure of aquatic communities under different degrees of pollution.

which identify the maximum beneficial uses to be made. From data so obtained the system's assimilative capacity can be derived, thus providing information about the investment in waste treatment facilities; the identification of valuable and unique wildlife resources; the formulation of a plan that will minimize the ecological impact yet allow maximum beneficial use of the receiving capacity without damage to the system.

- It is possible that during a pre-construction survey, a group of trained personnel might determine that certain areas of a stream were being affected by existing discharges or other adverse conditions, both man-made and natural. This is important information since it would be difficult to separate the effects of natural and man-made stress upon a stream. An example of natural stress was observed by one of the authors along the Texas Gulf Coast on the Sabine River. Every year during low river flows, a salt water wedge moves up the river. This wedge is destructive to fresh water bottom fauna and fish inhabiting the river's lower reaches. Only through a pre-construction stream survey could the effects of the wedge be studied and documented, thus protecting a paper-mill against blame for damage caused by the salt water invasion.
- Pre-construction surveys may dispel or confirm some beliefs of local inhabitants about fish productivity of a particular stream or lake. For instance, poor fish productivity in reservoirs may be blamed on industrial discharge when actually it is related to the aging process and is a completely natural phenomenon.
- Pre-construction surveys may indicate ways of eliminating problems before they occur. An example might be the placement of a discharge pipe so that the effluent would not endanger a popular fishing spot or the spawning area of a particular species. In other words, minor changes made before construction might eliminate major headaches after construction.
- Pre-construction surveys lend much credibility to statements by industrialists that they want to protect our streams and rivers. A company that has commissioned well-trained investigators to determine possible damage to a river receiving system, and has followed their recommendations to avoid it, is less likely to be condemned by government agencies and conservationists as lacking concern about that possible damage. Most industries discharging waste into the environment produce some measurable ecological effects. Few, however, have data which can be used to determine the severity of damage and thus make factual rebuttals to charges levied against them. Pre- and post-construction surveys provide this kind of information.

Post-construction surveys

The needs for post-construction surveys are:

- To determine to what degree (if at all) a plant's waste discharge has degraded the receiving stream.
- To determine accumulative effects of a waste upon aquatic organisms. For this reason post-construction surveys should continue as long as an industry discharges potential pollutants into a stream – once the survey is discontinued, accurate statements can no longer be made about the biological, chemical, and physical condition of a river.
- To demonstrate to the public that an industry is truly interested in maintaining a healthy environment.

Periodic post-construction surveys (at least once a year) are necessary even if an industry relies upon biological or chemical monitors to give continuous information about water quality in a stream. These instruments measure only a few critical parameters. Therefore, monitoring is comparable to having one's temperature taken – a survey to having a complete physical. Only through an intensive survey can investigators determine if the data from the monitors fall within an acceptable range. If one is assessing the impact of a waste upon a complex receiving system one must regularly look at the entire system in all its complexity. There is no other way to acquire a strong information base permitting accurate statements about the environmental impact of a waste discharge.

It should be understood that these intensive surveys will be moderately expensive. However, the services of a well-trained team in generating a broad data base will be far less expensive than costly court cases, adverse publicity, or waste treatment facilities that for some reason are not doing their job. The exact cost would depend upon a number of variables; but the contributions made by pre-and post-construction surveys to the designing engineers, the industrial public relations man, management, stockholders, and the general public far outweigh any costs that will develop.

BIOLOGICAL ASSESSMENT OF POLLUTION

In the past, biologists were more interested in the *kinds* of species present; now they are usually more interested in the *relationships* between organisms and their environment as well as between the organisms themselves. These relationships result in communities whose structure seems determined by the energy or nutrient flow through the network of species, often called a "food web."

Although the number of individuals at any one energy level may vary, the number of species playing a significant role in the web or community remains relatively constant under normal conditions. The number of components or species in this energy system also remains remarkably constant from one river basin to the next, and at various points within a basin even though the kinds of species comprising the system may differ at various sampling points. Thus, a detailed study of the miscellaneous collection of aquatic organisms in each receiving stream is not necessary; rather, the focus is now on the basic changes occurring in the system itself.

Most forms of stress (or pollution) reduce complexity. From a biological viewpoint this is expressed by a marked reduction in species diversity (or the number of established species present). However, the number of individuals may not change since the species able to tolerate the stress may

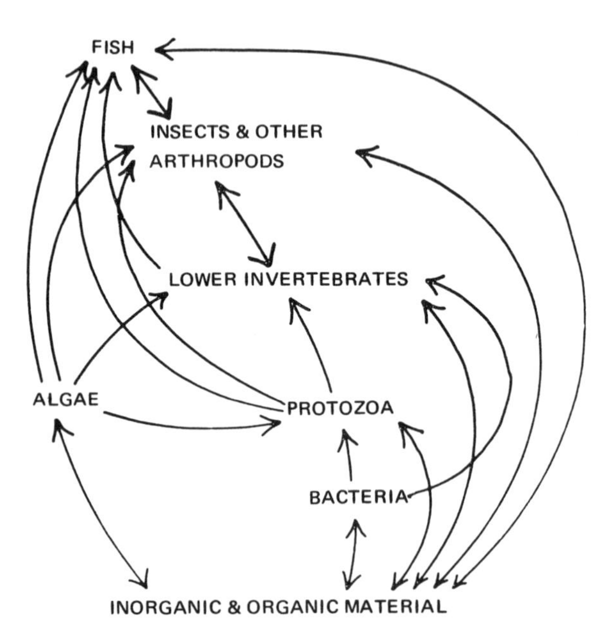

A simplified aquatic food web.

expand to fill the vacated niches. This might be acceptable were it not that simplification increases instability. The reason is that a constantly changing environment will probably affect only a small portion of a complex community at any one time, whereas half the individuals in a simplified community (consisting of a few species) might easily be wiped out when conditions become unfavorable for only one species.

This is, of course, a generalized summary of a complex situation. One final point on the biological assessment of pollution – it is important to recognize that biological data does not replace chemical data, nor may chemical data adequately replace biological data. They provide converging lines of evidence which supplement each other, but are not mutually exclusive. We are, hopefully, past the era of simplistic solutions to the complex problems of water management.

AUTHOR INDEX

KEY-WORD TITLE INDEX